LEANING ON A GATE

Published by

Librario Publishing Ltd

ISBN: 1-904440-62-2

Copies can be ordered via the Internet
www.librario.com

or from:

Brough House, Milton Brodie, Kinloss
Moray IV36 2UA
Tel /Fax No 00 44 (0)1343 850 617

Printed and bound by Digisource GB Ltd

LEANING ON A GATE

ELIZABETH MACPHERSON

ILLUSTRATIONS BY EVA SKEA

Librario

INTRODUCTION

Bob Geldof and the Boomtown Rats weren't the only people who hated Mondays. This was the day our mother chose to cease being a mother and wife, and boy, did we hate that. Monday was the day she became herself by writing newspaper articles recording the everyday life on an upland Morayshire farm. She did this fifty-two weeks of the year for over thirty years. Most of her output was printed in the *The Glasgow Herald* and *The Bulletin and Scots Pictorial.*

This book is a selection of her writing which records the changing scenes and patterns of countryside life spanning the years from 1945 to 1960. She had a country woman's eye for the small details and happenings which add much to life in so remote an area. We see the land-girl, Smithy and the Italian prisoner of war, Frank, playing their crucial part in our recently widowed mother's life on the farm. The grieve, Dod, Geordie are one and the same person. He too played his part on the farm and in our mother's life by marrying her.

The first passage is the last article her husband and our father, Ian Macpherson, wrote before his fatal accident in July 1944. He was also an author and broadcaster of great skill whose novels are rightly regarded as an important contribution to Scottish twentieth century literature.

"Leaning on a Gate" records the voice of an academic, humorous, warm, inquiring country woman who saw beauty in everyday things and was blessed with the ability to write about them so wonderfully.

Jane Yeadon
Elizabeth Baillie

March 2005

The Author, Elizabeth Macpherson, at her typewriter.

COUNTRYMAN'S CRACK

REWARD OF HUMILITY

(This was the last of Ian Macpherson's writings, received in the post just after the news of his death in a motor-cycling accident near Forres)

Some time ago I was sowing manure on young grass with the aid of an old, noisy bone davy – manure distributor to you – and an even more ancient and noisy tractor, whose engine roared and whose gears howled in deafening style. Yet as I went loudly round and round I found myself listening not to the mechanical din but to the whisper of rye grass heads rustling on the underside of the distributor, and to the soft patter of the manure as it fell in a meagre shower on the ground.

It was astonishing to discover that I could hear any other sounds than the iron clamour in which I sat enthroned; it was most astonishing to find that the murmur of the grass, the faint rustle of the manure falling a few inches to the ground, was more noticeable and more persistent in my head than the yells of the antique tractor.

But the small things are often thus noticeable quite out of proportion to their size. This year we had the biggest blaze of broom and whin I ever saw. Every bush by every road in all our country round might have served for a new sign to a new Moses, flame unconsumed among the hills. Yet many a time when I went out to see to my cattle or crops I found my eyes fixed on the ground where were tiny little flowers and herby-smelling things scarcely daring to call themselves flowers.

There grows here in waste bits of ground and on heathy land a minute yellow flower which country people call the darn floo'er, because it is said to have a binding effect on sheep. It grows in plenty right among the whins and broom, and many a time I have walked through the gaudy whins with eyes for nothing except this ground-set weed, this tiny little yellow-flowered thing which you can't see from five yards off. Is it because these flowers are small and weak and yet dare the eye of day and grow and serve their own lowly purpose?

Perfection is seldom large, but resides most commonly in small things like wrens, and the smell of herbs, and small sounds. Mountain and great seas, and man's own proud works, too, may excite the aspiring mind and convince you of the glory and wonder of the world which is as fair as dangerous. You are more likely to find utter satisfaction in the loveliness of the earth when you look down at the things you tread on daily. Humility has its own grace and is not without its own reward.

SPRING

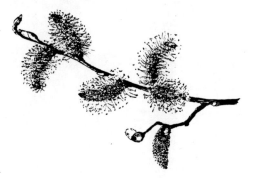

ONE DUCK – AGED

When the agricultural returns come round, that is how we describe our solitary Aylesbury duck. She has been appearing thus among officialdom these many moons and wears her years with intelligence and not a little dignity.

We gravely worked out her age the other day and have come to the conclusion that she is 15. I am happy to say she looks good for 15 more.

Duck leads a solemn and circumspect life. By day she keeps company with two hens and an old cock.

Long before the hens can get near a worm, Duck's flat and predatory bill is gobbling up the delicacy. At half-past twelve midday she comes to the back gate, demands a warm tattie. It must be warm but not too hot.

Should there be what Duck feels an undue delay she swears something horrid. The Cock who is a well-mannered old gent remonstrates with her and the things she says to him about his gout and his grandfather are so obstreperous that I don't wonder the two hens blush right over their combs.

Twice a year she lays her eggs, large and white in impossibly secret places, and twice a year we organise efficient and patient search parties to find them.

After each find Duck sulks angrily and after the autumn find gets so mad she takes off practically all her clothes. Even she is a trifle ashamed of this hysterical abandon so she wears a puff of down on top of the accumulated mud on her bill and trusts that no one will notice the paucity behind.

But life is not always thus even and quaker coloured. As soon as the sun sets she escorts her daytime companions to their house and having seen them safely perched she departs in search of divertissement.

Sometimes in the deep dark you can hear the greedy gutter of her questing bill in nameless puddles and pits. At other times she pretends she is sailing on some wild and distant tarn when all that lies beside her is my old enamel basin full of water.

But best of all, she loves a night when the moon is staring white and full. Then, all alone she dances and singing cheers the stars on to paradise. No longer fat and wheezy, she feels herself flying along the arcs of beauty down the night.

And humans, hearing her raucous shouts, mistake her immortal longings and merely say "Hear Duck! It'll be wind to-morrow."

SPOT AND TINY

Naturally having sworn after last year's searing experience with a tame sheep which started life as a pet lamb, never to have another sheep about the place, we find ourselves this year not with one pet lamb but two. Spot arrived first, a present for the children from the same kind friend who gave us Christina last year. In spite of the fact that I had known that holy terror intimately, I fell once again for her self same charms now embodied in the white and innocent Spot.

Sighing at my own inconstancy I went off to search for Christina's old rubber teat to give Spot his first meal in his new home.

It was very difficult hardening one's heart to the blitheness and trustfulness of the new arrival, but I felt in self-defence I'd have to do it.

I was getting on grand and managed to feel fine and cynical at the sight of the lamb's small head resting confidingly on the spaniel's black flank while the two of them slept in the sun. And then Tiny appeared one wet Saturday afternoon.

The grieve found her lost on the hill weeping her sad heart out. There was no sign of her mother anywhere, so Tiny was borne home and please could she have something to eat now – baa-baa-baa.

So we warmed her and dried her and fed her, and hastily sent word to our neighbours that we had found a day-old lamb, and would whoever was short in the count please come and collect same.

In the meantime we would continue our hospitality, and Tiny could take her chair right up to the fire. And when she did this with the greatest alacrity we said, "Och, well, it won't be for long." But the days passed and Tiny remained unclaimed. She is still here.

She is extremely pretty, being a Blackface lamb and not a cross specimen like Spot, which loves her to distraction. This pleases Tiny very much, as she is a hussy of the most coquettish.

Why, would you believe it, the wee limmer, not content with the long, lashed amber eyes that nature gave her actually has mascaraed them. Frightfully fetching, especially as the cosmetic seems to have slipped a bit and adorns her greedy little nose as well.

When she came first she came the poor, pathetic, clinging-vine stuff, but as she has grown in size and worldly wisdom she now adopts the tomboy with the romantic past attitude.

This pose she finds extremely useful as it allows her to steal the hens' food with gaiety and also escape the kicks and cusses she deserves. Who could ill use an unknown orphan?

Besides her folk might be anybody. Looking at Tiny and Spot I take a dim view of the future of this farm.

A PIECE DINNER AND TEA IN A FLASK

4 March 1961

This is the time when we are busy with the spring mills, which we hope will give us a decent sample of our own corn for sowing. Because last year's harvest was so hazardous many upland farmers for the first time in their lives had to rely on the combine to save a proportion of grain. The result is that our cornyards are half full and that, coupled with growing caution about cereal production, means that, instead of the usual whole day of the mill, half a day is enough, and even in some cases a few hours.

Therefore the neighbours who have been lending a hand stream home by car at midday and the farmer's wife, who for years has been accustomed to monumental catering on a mill day, sits down with a shock of surprise to a dinner the same size as yesterday's. She can hardly believe that the mill dinner is finally over; no longer need she worry about bad weather cancelling the whole complicated enterprise and forcing herself and her family to eat their way out of the gallons of broth and the gigots of mutton which were to have fed a score of hungry helpers.

Actually the mill dinner has been on the way out for some time. Farmers who came to lend a hand on a mill day came by car, and, since they had nobody at home, they used to dash back at dinner time to sort their beasts. By car it was easy to do this and it was only common sense to take their dinner at their own tables. If they also had a worker helping at the mill, he too went home, either in the boss's car or on a tractor, so that only the mill men and one or two folk from very outlying places were guests. Of course there were still the morning and afternoon teas, but they were a mere nothing.

No More Ulcers

But the countryside is not tenanted entirely by farmers and their workers. Roadmen, forestry workers, railwaymen make up a good part of our rural population. Seeing how other folk got home for dinner they were not long in following their example. Lacking the tractor and car they fell back on the useful little auto cycle, so that now you see everyone who can quietly "put-putting" his way home at midday. No more ulcers from the "piece dinner"; no more headaches for the housewife who had to try to vary the contents of the dinner-box as much as she could.

Of course not every country worker can make such an arrangement. Where he is employed by a contractor who supplies a lorry for transporting his workers to the scene of their labours he must accompany his mates, in which case he still carries a "piece dinner" and the inevitable flask. I often think if I had money to invest I'd buy shares in a vacuum flask concern, since the casualties must be enormous.

A number of Forres men work at the aerodrome at Kinloss and as they are civilians they make their own transport and meal arrangements. They are all friendly chaps and, since they form a small enclave on their own, they take much interest in each other. That is why the other week when one of them threw away his sandwiches to the gulls with a disgusted snort "Och me, cheese again" his friends were concerned. He was offered pieces from other boxes, but the cheese, he said, had "fair seeckened him."

Next day the same thing happened. "Cheese, aye cheese! Ach well, the gows can get it". When this had gone on for a week his worried workmates tackled him. "Whit wey day ye no tell yer wife ye dinna like cheese" they asked. "I hinna got a wife" he replied. "Weel, yer mither" they suggested. "Ma mither's deid" answered the disgusted one. "Fa makes yer piece then?" asked the perplexed friends – to which he replied savagely "Masel."

DISCONTENT WITH WINTER'S DUST

11 March 1961

All the hours of spring sunshine pouring through the farmhouse windows have the sad effect of highlighting the discontent of winter's dust; and even if farmers are engrossed in the unusually early spring work their wives decide that for once the annual cleaning cannot wait till the corn is sown or the clean ground ploughed. We make for the telephone before our husbands beat us to it, but sadly find out that all the tradesmen we want are already fully occupied in hotels and boarding houses. Every year our district becomes more tourist-minded, and those of us who live on the periphery of the summer season discover to our deflation that we are no longer the chief industry of the countryside. Nor can we compete in gaiety nor finance with summer visitors.

Huge Buses

Meanwhile holiday traffic thickens daily along our hill road. Cars with caravans negotiate our extravagant bends and wild gradients on their way to hills or shore. Already at the week-ends we see the huge buses the airmen at the coast aerodrome charter for their weekly leave come cautiously round the nasty corner at the little bridge before facing the long pull to the summit at Dava.

Still the sun shines, filling the sky with exhilarating sharp light. We are so high here that the wind remains stimulating notwithstanding the real warmth in the air. The fire of crocuses flickers in the borders of our little hilly gardens, while scyllas and daffodils jostle beneath the sitting-room window. If one could count on this weather at this time of year what a revolution it would cause in our holiday habits, for no one could possibly remain at desk or factory while the compelling sun floods the hill and firths break their gentle waves against a sandy shore. Even I want to go and live for a day or two in a caravan on Nairn beach, though I loathe being infiltrated by sand which does not stop at shoes but works its gritty way via pots and pans into food and probably into beds.

However, the sandy van, the air tasting of salt and iodine, and the clear, high call of seagulls must remain purely academic as far as we are concerned. Holidays for farmers are chimeras. When the weather is bad you want to stay at home by the fire, and when it is good you are seized by an obsessive compulsion to work. If there are no calves or lambs needing attention, there are crops to sow or garner. Even if by some triumphant magic you could arrange for all the cows to graze peacefully on summer hills, while all the grass refused to pass into useless maturity but stayed at the right stage for hay or silage, there would still be the difficulty of the household pets.

It is useless on a farm to harden your heart against animals. All the same, no landlady is going to welcome the large woolly sheep who was once a bottle-fed lamb and who now insists on accompanying you as if you were her mother. I quite realise that a ferret in one's luggage will attract unfavourable comment. I cannot think why Dod and the girls are so foolish about ferrets, who strike me as unendearing and smelly.

Bob the Dog

I suppose we might manage Bob the dog. Perhaps if we insinuated that he could do a sheep dog trial act for other guests we might wangle him in. But on second thoughts I doubt it. We once took him for a holiday down to the shore where we happen to own a house. We motored down and Bob said he felt travel sick, so we opened the windows. Immediately he tried to leap out and organise some sheep belonging to another farmer. Finally he leaned his weary head against Dod and slavered nervously down his neck. Our holiday was correspondingly nervewracked, so I think we'll just stay at home as usual.

SETTING THE HEATHER ON FIRE

18 March 1961

The spring morning smiles delightedly on every little rolling Morayshire hill where, even at this early hour, shepherds and keepers are hard at it burning heather. The season for this operation is a short one and if it is to be used to advantage you must make use of every single puff of favourable wind.

For the past week we have been much at the caprice of a breeze which begins sweetly and gently with morning and then dies down completely at noon. For an hour or so there is calm and all of a sudden a gale gets up and from a new direction with alarming consequences. One of our near neighbours had a very frightening experience when this happened to him a day or so ago.

Wall of Flame

He turned his back on a fire he imagined he had extinguished but the malicious wind blew it up again and before he knew it a huge wall of flame was advancing towards the big plantation which separates his ground from the next estate. He had to get fire engines and squads of men to deal with the emergency. Now he looks with horrid misgiving on the amount of money he must spend on renewing fences devoured by the voracious fire. The rest of us run to check our fire insurances in case the same thing happens to us.

But if one forgets the danger, heather burning can be a lovely sight, with long blue plumes of smoke rising up against the clear white light flooding our sky in spring. Pigeons soar and veer above the moor and the sun catches their white breasts so that they look not birds but stars. The hill itself still wears the drab of its winter herbage but flames, orange and red, grow out of it like flowers. At night we can sometimes see heather fires burning on the hills across the firth, bonfires in honour of spring.

Neglected

Heather burning has for long been neglected in hill management. For a time we used to blame the war years for the waist-high scrub that disfigured and made useless so much of our high moors. But actually the neglect went much farther back – to the days when the sporting estate was a major factor in the economy of the Highlands. Landlords then believed that long heather was essential to the preservation of the grouse. Sheepmen, on the other hand, took the opposite view and complained that the failure to burn the hill was ruining its feeding value. Continual bickering did not help, although eventually some lairds did come to see that sheep and grouse were not the natural enemies they had once thought.

Now the great feudal ramifications of the season and the sporting estate are over, and the landlords begin to see in their hills another though less picturesque source of revenue. Round about us more and more Highland lairds are going in for flocks and herds reared on the summer hill and wintered at home in the sheltered arable farms. The modern pastoral laird is something new in Highland history. He combines an eye for a beast with a word-perfect knowledge of the Marginal Agricultural Production (Scotland) Schemes and the Hill Farming and Livestock Rearing Acts.

Somewhere the old Highland minister, Donald Sage, speaks of "high-souled entry, enlightened by divine truth and knowing their Bibles well." I could not say myself how far divine truth has been responsible for the lairds' acceptance of the necessity for liming, draining, and burning the hill; but there is no doubt as to the whole-heartedness of their conversion to sound hill management.

The heather fires about us leap and lowe under the wary eye and birch besoms of the men who control them, and I wonder if after all the talk about regenerating the Highlands this may not be the answer.

THE AIR IS FULL OF NOISES

19 March 1945

Yesterday afternoon I went down to burn the straw that had happed a tattie pit. This is not so easy as it sounds, because the straw was wet, and, moreover, it was half-buried in earth. However, with the aid of some dirty paraffin I got on not so badly. As I howked and peched and gathered odds and ends of dirty bunches, I had time to notice all the clamour which is spring's accompaniment.

The wind blew loudly in the ash behind me and made my fires roar.

Have you ever noticed the differences in the sound of the wind as it blows through different trees? In beeches its loudest note is still a sussura. In the ash and oak it sings like an organ, but in larches it cries like something lost. In a thin belt of spruce it can shriek, but in a deep wood of pines its wildest fury is a lullaby.

Above my head the larks poured out their rapture, and curlews mocked their happiness with their two-toned call of grief. Not till evening would we hear the sweetly suffocated singing of the thrush and the carelessly finished golden treasure that the blackbird tosses to the first star.

In the meantime an eddying cloud of gulls were trying to outscream the whee whee whee of the Ford Ferguson which was relentlessly throwing up furrow after furrow. An indignant grouse forgot the time, and ko-axed furiously at the world from a hillock on the moor.

Away at home I could hear a fine spring sonata. A dog barked incessantly, either rebuking a stranger or chasing a rabbit. A steady bellow borne on the wind told me that little fat Kate was looking for one of her calves. A neighbour living three miles away avers that when Kate is in voice he can hear her quite clearly. I can nearly believe him. If Kate would only alter her note I could bear it better, but she keeps up a steady "ba-ba" that deaves you. This afternoon she and the Fordson were having a duet.

The Italians were using the Fordson to cart dung, and I knew by the sound that Frank was nursing her like a mother over the cobbles of the square. I sent up a small prayer that when I returned he would not confront me with fresh evidences of my tractor neglect.

Lorries laden with straw and hay and all the rest of the spring carting laboured up the hill. The stots rustled in the broom.

And then came a final blast of wind. It died down and my straw was no more. Grey dust blowing over the furrows told of a year that was gone. The wind, rejoicing, spoke of another year.

PUBLIC RELATIONS

20 March 1951

Of late there has grown up among Government Departments and large industrial concerns the habit of appointing a public relations officer whose job it is to explain the motives and aims of the corporation he serves, and as a corollary, flatter the public into thinking everything is done for the sake of the man in the street. I could wish that farming employed such a one.

I don't think that farmers as a class are exactly the blue-eyed heroes of the populace. And this is quite understandable, for the deep subconscious of every citizen is certain that food ought by rights to be laid on like air.

After all, don't we all in the beginning come into a world where beneficent Mother Nature provides a convenient supply all our own?

Unfortunately this agreeable amenity does not last long, and in fact the procuring of food gets progressively less convenient until we not only have to find it and cook it, but pay for the stuff as well. And the man who is responsible for this last insult is a farmer.

Farmers are very conscious that they enjoy initial disfavour with the rest of the country, and they do try to counteract this antipathy in a number of ways.

Simple souls, they believe that if the townsman but knew about farming he'd like the farmer. So the farmers arrange such excellent shows as "The Country Comes to Town" where they demonstrate their way of life.

These shows are enormously popular, and though the impressive display of technical and scientific achievement necessary to run even a little wee farmie passes the urban dweller by, at least he does learn that cows do not produce milk all ready in sterilised tin-foil-topped glass bottles.

But much more than this is necessary if the farmer is going to sell agriculture. He can best realise the magnitude of the task by comparing the publicity any other great industry received with that accorded his own.

"Ten million pounds for the Farmer" scream the headlines and every one believes it except the farmer, who knows that this sum will in reality be so hedged with restrictions that it'll be a wonder if one farmer in a hundred benefits by it. No newspaper equally loudly shouts "Millions for the Miners."

Has the fact that every farm works seven days a week any news value at all? No, but Parliament itself records that so many pits were open on a Saturday afternoon.

Have you ever heard the BBC announce, all bated blandness, that the farms of Britain have produced so many million gallons of milk for the week ending such a date? This represents an increase of 32 pints on the corresponding period last year!

What happens when the farmer objects to the call up of Z men? "Listen to the farmer squealing again" runs a letter to the editor but "The country needs the Z men in the mines."

Did I say agriculture needed a PRO? It does, but only the archangel Gabriel has the qualifications.

SPRING FEVER

25 March 1948

It is spring and we are all busy. At one moment we seemed to be living in a meditative countryside where snowdrops bloomed in drifts and cascades and only the bright sun shining in the forest aisles was alive with quickened purpose.

But overnight it has changed and no one has as much as a glance to throw on the sunny, gusty landscape save to count how many more days it will take to get the stubble ploughed out.

In every field, tractors caterpillar their indefatigable way, adding to the brown and winnowing furrows. The snell wind blows, the infrequent gulls lean on the bright air, and the tractorman, hunched between his tractor and following implement, calculates to a nicety his finishes.

On the roads heavy lorries snore up the hill to rumble into farm steadings.

The drivers enliven the job of unloading the spring manures with fabulous tales of how much further on the work is down in the Laich than up here.

If the stories of tens of acres of tatties planted in a day and all the grain crops already in weren't so miraculous we'd all be green with envy, but signing the nitro chalk chit we reflect with a pleased grin that for once our own work isn't too badly behind.

Indeed, when I remember that on this farm last year by March 20 we were still without a single furrow turned, I am more than grateful to 1948, which has let us get on so well.

There are, of course, the usual complicated and nerve-wracking arrangements to be made by telephone anent things like seed corn, mercurial dressing, and Government help in the shape of discs and grain drill and such like.

But all these fiddling details cease to be irritating and anxious when the grieve admits that three more days will see most of the black ground ploughed.

The beasts, too, seem to share in the general excitement and the cows go far afield searching for a fresh bite of green, while the little calves kick up their cheeky heels like bovine ballerinas.

The only person really keeping aloof from all the pleasant steer is Duck.

She, depraved bird, has lost an eye. I don't know what drunken brawl was the cause of her mischance, and she maintains an obstinate silence herself.

She contents herself with plowtering through her muddy ways morose as a dyspeptic pirate condemning all her victims to walk the plank.

FAITH, HOPE, AND CHARITY

25 March 1961

In another week the wintering sheep will be gone, and all we'll be left with will be our own three Leicester gimmers whom I call Faith, Hope, and Charity.

Dod acquired Faith, Hope, and Charity during one of these interminable, complicated transactions which are crystal clear to farmers but which drive economists into a frenzy and reduce the Inland Revenue to impotent nailbiting.

John is our neighbour two miles up the road, and his farm is predominantly a sheep one while our one is cattle raising. He finds it convenient then to send us his Leicester tups for the summer and we find it convenient to board them since they do our grass a great deal of good in that they keep down the ragwort. Ragwort, stinking-willie we call it, is poisonous to cattle and in silage it is deadly. Therefore we have to extirpate it as best we can, and up to now have found no herbicide half as good at the job as the close-grazing sheep, which find its young rosettes both palatable and nourishing.

John appreciates same and bawls cheery greetings to us as he passes on the activities which are appropriate to sheep farmers in the summer. Dod however begins to think the tups are not getting enough protein and he stops John the next time he sees him driving his car down the hill.

"Och, Jeepers," says John, not wishing to be hindered. I don't know what Dod takes this to mean, but next day he ostentatiously sets out feeding troughs for the tups and John comes zooming in with two collie dogs in the back of his Husky a powerful smell of dip and a box of sweeties for me.

The Bargain

John does not disapprove of Dod's methods; in fact if Dod were not to give the sheep the little extra John would find a way to tell him so, but the essence of the bargain is never put into words. Money is not mentioned.

Then I find that the hens' corn is done and I ask Dod to get some; note, not buy – get. He disappears and five hours later announces that John will bring down the corn a week hence when he comes to collect the stirks which Dod has sold him. I have not known that the stirks were to be sold at all, but I am so well trained I betray not the faintest surprise but only emit a gloomy mutter about the Price Review.

A day later Dod tells me that he fears he has driven a terribly bad bargain. After all, the stirks have not been punched and now John will collect the appropriate payment from the Government instead of us. The next time I meet John he comments on the appalling usury of Dod who just because the stirks are not punched has extorted vast sums from John. Naturally I believe that money has indeed been for once exchanged and suggest that as we are going to town Dod had better deposit same in the bank.

How wrong can I be? No money has changed hands. How could there? Neither Dod nor John have the time to sit down and calculate what each owes the other. "Och, Jeepers," There is the summering of the tups, there is the extra protein, there are the stirks less the headage, less the luckpenny, plus the extra week's keep at the hungry time of the year, less the transport on the hens' corn, less the two hands at the mill before last.

I trust that now you will see by what devious routes we entered into possession of Faith, Hope, and Charity.

TALE OF YESTERDAY

27 March 1951

The wind was grey out of the Knock and the ditches and drains by the roadside were full to the neck of gurly thaw water. Down in the little croft about half a mile from here, I found my old neighbour strong in the dignity of her eighty-odd years and a pair of tackety boots, dealing with flood water which was making a mess of her diminutive barn.

What prodigies of skill her nimble broom performed! With what incredible swiftness she pattered from one end of the building to another!

Her sedate white and black collie dog and myself surveyed all this energy with appropriate admiration but were truly thankful when her stern blue gaze allowed the matter of the barn floor to be well and thoroughly accomplished.

"I wis learned tae work when I wis young," she told us as she led us indoors. "I had tae work," she added grimly, as she changed her working boots for the gym shoes she affects as slippers.

"Ye see," she went on, "my faither died when ma youngest brither was three weeks auld. Aye, we had tae go on the parish then. They paid ma mither a shilling a week for each bairn. There wis fower of us. Ma mither gaed tae work and her first job was spreading mole heaps wi' a shovel. She was owre poor tae buy shune, so she worked bare fitted. Her pay was eightpence a day."

Outside the little window of the immaculate kitchen with its pipe-clayed fireplace and hearth where the peats glowed, the bitter spring girned in the irregular braeset fields. Would you not too have shivered at the quiet Scots voice telling of a woman working barefoot in the fields for eightpence a day?

"Oh, bit things did get better after a whiley," continued my old friend. "There wis a hairst and she got saxteen shillings in the season. The men that workit on the place were awful decent and so that ma mither had no broken time, for ye see she'd no get paid for broken time, they used tae tak her tae the barn for tae mak straw ropes.

"The way they made the straw ropes was wi' a thing called a thaw crook. It had a great long iron spike and was fastened roon the waist wi' rope. The men paid oot the straw like as they were putting wool on a spindle and syne the woman twisted and twisted till she's made a rope.

"They were graund ropes, the straw ropes. Strong they were. So they had tae be for they were used tae tie doon the stacks. Nae Glasgow Jock in those days.

"Weel, at the end o' that hairst ma mither got the full pey. When she saw the lot o'sillar she thocht it couldna be her that had gotten a' that."

From our window which looks into the brown moor you can see the old feerings which the encroaching heather is powerless to obliterate. Do these undying memorials to a peasantry of yesteryear still hear above the wind the voices of the barefoot women who once worked there? A hard life it was, but somehow triumphant.

BIRDS, BEASTS, AND WEATHER

3 April 1951

Since November we have never been entirely free of snow, and if we humans are tired of it so are the beasts and the birds. At this time of the year the farmyard beasts, the cattle and sheep, begin to yearn for the taste of fresh growing food as a change from the hay, straw, and turnips which have been the staples of their winter diet.

As a rule, too, although the growth on the hill is infinitesimal the beasts have been able to get fresh bites here and there, but this year the hill is still snow-bound, and instead of foraging widely as in other springs they hang disconsolately about the muddy gates.

On this farm we are thankful we have no sheep, for the lot of the sheepman with lambing ewes must be unenviable.

Birds, too, feel the inclemency of the time. We have a blackbird and his wife who live in the juniper bushes behind the house, and though they intend bliss, so far they approach it with a dance, but no song.

Mutely they play hide and seek in long swooping flights, and Mrs Blackbird tries not to droop too sadly on an upturned pail as she remembers the serenades of other springs.

Whenever the lashing storms of snow and sleet allow, the larks chorus rapturously, but, alas, how sadly soon the ever encroaching storms silences them.

The moor birds are as unhappy as the others. Here at least it has been quite impossible to burn as much as a tuft of heather, and the streaming hill won't make the best nesting.

We are still without curlews. Some weeks ago a few did venture up, but one look at the snow-covered hill was enough and they rushed off protesting they were not ptarmigan.

The peesweeps have behaved perceptibly more bravely. Several rather anguished flights of them are in residence and they can be heard mewing within the storm clouds with the eerie significance of the protesting wild.

But the bird for whom I am sorriest these cold days is my old duck. I fear she is at last paying the penalty for her debauched past.

Her feathers are deplorably ragged and she is sore afflicted with rheumatics in her right leg. All day she sits in the lithe of a shed nursing her poor cold feet and suffering shocking persecution from her fellows.

When I rescue her with pitying cries she clucks softly in her throat and returns to solacing her aging limbs and enjoying her memories.

What pictures flit within that tousled head? Whatever they may be, whether of summer in the water meadows or but dim ancestral recollections of fleeing the Arctic darkness freaked with the armour of the cohorts of light, they all vanish before a delicious present of warm wet mash.

With the hot food inside her she says she is sure that spring and sun will come at last. Dear comforting Duck.

APRIL FOOL WEATHER

8 April 1961

Most of the wintering sheep are home now, though there is still an anterin double-decker float lumbering down our hill roads with the very last of the returning flocks. Because the spring has turned so cold and dirty, we are thankful that we saw the back of our winterers a week ago. We were lucky too, in our gamble with the weather. We were sure that it would take advantage of its being the first of April to try to make fools of us, and it did its best.

The previous dry spell broke neatly and thoroughly the night before so that the farm was in a lovely muddy state when the first heavy lorry came in. But the joke was on the weather clerk since we had laid down a good thick mat of gas clinkers in the approach to the loading pen as well as in the pen itself, so that there was hardly any bogging, and the only real hindrance came from slight arithmetical discrepancies between the counters.

I cannot imagine why counting imaginary sheep should be thought as soporific when counting real ones can be such a headache. A score of Blackface ewe hoggs going like the hammers up a ramp and into a truck takes some distinguishing – the swiftness of the hoof deceives the eye. When you have 300 of the brutes the confusion is by that increased. I like the story of the old shepherd who counted his sheep in tens and at each ten he unloosed a waistcoat button. When he'd used up his waistcoat buttons he began on other ones and peasant humour can only too easily be turned to earthy laughter when it contemplates what happened when all the poor old gentleman's buttons had been used up as an abacus.

Angry Squall

As the float departed, an angry squall followed them to emphasise the fact that spring had not nearly got over her tantrums. If it had not been that hopeful winter sporters were already making their way to Grantown with their skis lashed to their car roofs, we should doubtless have had a right royal storm. As it was we had just enough snow to put paid to any effective work on the land for the next four days, but not nearly enough for hoteliers wanting to emulate Switzerland.

There is nothing more thwarting in farming than being pulled up with a round turn by the weather, but at least you can take advantage of the hiatus to bring your reading up to date. By reading I do not mean things like Critical Approaches to Eng Lit, but engrossing wee bookies about Field Trials conducted by your nearest College of Agriculture.

Engrossed in percentages of resultant dry matter and crude protein you "fleet the hours away as in a golden world" and never notice the east wind whitening the arid sides of the old Knock of Braemoray with snow showers.

Peesweeps making their tardy if ill chosen advent to the hills flap bravely against the cohorts of marching hail and scream "like fiends in a cloud." You hear them not. Should you chance the cheaper Westernwolths grass for a one-year silage? What delicious results can you expect from an additional two cwt of sulphate of ammonia to the acre?

Regardless of a landscape as verdant as the polar ice cap, you juggle with your cheque book and by the time you have mortgaged your future for years ahead, the weather looks like clearing. Manoeuvred once more into April folly there is nothing for it but to acknowledge that "age cannot wither nor custom stale the infinite variety" of Scotland's April.

OPERATION OSTRICH

13 April 1948

I never go to a mart but I am struck afresh by the invincible powers of the auctioneer who accepts a bid. You see, the main thing so far as the bidding farmer is concerned is that no one save his Maker (perhaps) and the auctioneer must know that he so much as desires to bid, far less buy.

The result is that every man devises for himself extraordinary furtive mannerisms which the eagle-eyed auctioneer must recognise. Good heavens! There are moments when I look round the ring of bidders and see my nice hearty farmer friends apparently afflicted with incipient epilepsy.

One man rolls his eyes to heaven, another winks frantically, and a third swallows his Adam's apple with alarming rapidity.

Alec doffs his cap to wipe the sweat from his fevered brow and signify that he advances a couple of quid on Rob, who ostentatiously stubs his cigarette to show he doesn't want the damn beast – not at that trade anyhow. It's all very entertaining but a thought bewildering.

A shrewd story was told me the other day by a friend who is a farmer's wife. There was a sale of furniture which she wanted to go to because there was, among other things, a good dinner set. Being a dutiful and comparatively admiring wife, she asked her lord to come and bid for the dishes. After all weren't auction sales his particular meat? Well then. So they went.

The farmer watched excitable women bidding loudly and raucously. Contempt lined his rugged features. Came the dinner set. He prepared the weapons of his extensive armoury.

He pulled his cap over one eye, he pretended he was measuring the middle distance with the knob of his stick, and by the time he had wagged the left leg of his horn-rimmed specs the dinner set was knocked down to a cheerful Bacchante who had merely bellowed louder than her fellows and waved a flamboyant umbrella.

The farmer gave an outraged gargle and stalked from the irresponsible scene. His wife followed – in silence – more clamant than the brazen trumpet.

But the sappy bit of the whole thing is that although the farmer acts like the proverbial Sphinx in the sale ring, yet his real business, which lies in his fields, is open for the whole countryside to see, comment on, and criticise. Which it does – with relish!

LEISURABLE HOURS

15 April 1961

Walking for pleasure today seems to have vanished with many other simple, uncomplicated recreations. When I was young, the hiker was a familiar figure on every country road, but it must be years since I have seen his bowed figure with an enormous pack on his back which made him look like the illustrations in my penny edition of the Pilgrim's Progress – only I do not recollect Christian wearing such impressively hob-nailed boots. The young folk today still enjoy going on cheap adventurous holidays, but they seem to go by bike, or scooter, or even bubble car. No one walks.

Hill Track

But I still like a mindless dander with no purpose but to enjoy the hill air and appreciate the view. When I want to do this I go up the hill track we use for driving the cattle to the moor, and turn sharp right over a mouldering gate and through a disintegrating fence into the planting which traverses the rising moor above the road going south to Grantown. This planting has recently been thinned so that there are pleasant convenient tracks all ready for some one who likes the illusion of rugged privacy without the drawbacks so attendant on walks in any upland country. Wearing rubber boots and leaping like a hind among boulders and hill drains is all very well, but it lacks the serenity which is the great charm of a walk through the wood.

Cut off from the world within the piney retreat you may wander as freely as you will, secure in the knowledge that all you will meet is the startled capercailzie or the shy roe deer, both of whom are as unwilling to meet you as you are to encounter the great lorries rumbling down the invisible road a hundred feet below. Woods have a wonderfully tranquillising effect on winds and noises.

The path through the wood is barely a mile long, and when you emerge at the southern end you find yourself in April sunshine but effectively sheltered from the April wind, with the great empty expanse of Dava Moor in front of you. A brisk burn dashes past, chattering excitedly as it runs on its way to its rendezvous with the Dorback rising away yonder out of grey Lochindorb. You are looking out over one of the routes by which an exasperated Malcolm IV is supposed to have expelled the quarrelsome indigenous men of Moray, who "cherished their warlike habits as if they constituted a moral code of infallible excellence."

Drove Road

The present road is, of course, a newcomer to the ancient magnificence of the scene. If you'd like to see an older one, just climb up the hill and you will presently be able to pick out the old track which was the drove road to Falkirk and other cattle trysts. The farmhouses bordering its verge are in ruins, but somehow they do not look all that desolate for a neighbour still cultivates the secluded acres, and green fodder crops make the prospect pretty and cared for.

Morayshire has a comforting way with history for she manages to emphasise its kindly continuity rather than its transience. The wood behind you sheltering you from the blast is part of Darnaway's huge timber holdings which had their origin in the fourteenth century. The name of the parish where we are means in Gaelic, "the face of the wood."

Old Vellum

In Elgin library, they can show you a book written by the only great philosophical scholar the province of Moray ever produced. Florentious Volusenus, born in 1504, wrote the "Dialogus de Animi Tranquillitate" for a recreation in his "leisurable hours." Looking at the antique vellum, admiring the beautifully printed Italian, one cannot help thinking had Volusenus lived today he would still have written in the same spirit.

FULL TIME WITH THE LAMBS

20 April 1948

We have no sheep ourselves but all our neighbours have, so they are busy night and day·just now with the new lambs.

The weather has been kind and the winter open, but for all that April is a hungry enough month in these upland regions. The winds blowing over Dava are too bracing to encourage such shy grass as we already have.

Indeed we must wait for nearly a month before we can emulate the soft thick emerald that now clothes all the fields and road verges in Forres 800ft down the hill.

Yet although we have no blossoming trees nor verdure we are not without the signs and affections of rainbowed April. And the lambs are the loveliest of them.

A little way down the road a neighbour has a field of turnips growing rich green crops, and here he has his ewes and lambs. Bales of straw in the centre of the field make a rude shelter for the blackfaced sheep and their dancing offspring.

To-day I saw a grave-faced matron lying with her back to the blast which was blowing her thick grey wool into fankles. Under her chin, cuddled up to her breast, was her new youngling, white as milk. And as she bent her quiet head above the little creature you had a sudden glimpse of all the love and patience which flowers for ever along the world's dark highways.

Winds wet with April swept along the shoulders of the Knock of Moray and the lamb secure within it's mother's care stared out with unbelieving eyes.

By the end of the month all the worries and triumphs of the

lambing will be over and as the hill grows green again under its blackened scars of spring burning the sheep and their lambs will go out to the heights.

Night's Sleep

As the shepherd watches the white flood going out through the hill gate he can look on them with gladness and relief and think with smooth pleasure that for the first time for weeks he can look forward to a full night's sleep.

He hangs his bag and lantern up behind the door to signify an end to another lambing while his wife shoos the orphan lambs, who for the past 10 days or so have been turning her kitchen into a sheep fold, out on to the greening patch of grass beyond the door.

Lambs, like children, love company and their new-found solitude makes them bleat in bewilderment. But they soon forget their unease and come running for their bottle of milk playful, exigent, and hungry.

COUNTING COUNTRY HEADS

22 April 1961

All this week three of us have been going round our curiously shaped parish delivering census papers, and incidentally brushing up our map reading and our stores of odd social knowledge.

My district lies on both sides of a back road in upland Morayshire. On an April day of sun and warm breeze there can be few more delicious ways of passing an afternoon than dodging out and in to farms and cottar houses. Birch woods new come to leaf, gardens shining with aubretia and primula, alternate with the long slow acres sown to corn. Down across the Laich of Moray where forward weather had already brought blossom and braird, you can glimpse the calm firth laving the feet of hyacinthine hills on the farther shore.

Camps of Tinkers

In comparison with the town enumerator our task is a mere nothing. · In the town each enumerator expects to deliver anything from 280 to 300 forms. All the addresses I was given came to 32, and of these six were of houses now vanished. Indeed there were two of them of which I had never heard so I rang up an old friend whose family have farmed in the district for hundreds of years.

Oh yes, he recollected the places very well, but today they were become farm buildings. He added that I would have a very easy time of it. The last census in which he had taken part had been held when there were camps of tinkers living by the bridge over the Findhorn. They provided colour and headaches for the conscientious enumerator. Their attitude to the printed word was carefree to say the least, and they walked off to other greener places bang in the middle of the form filling, joyously drinking whisky as they went.

Today we are so conditioned to schedules that we are much more amenable, and in rural areas we regard schedules as another occupational hazard like the weather. All my households took their forms as to the manner born, before going on to discuss things that really interested us, such as the feeding of draff to wintering cows and the comparative economics of the September and the March calf. We brooded on how much money it takes to produce good grass and we all took a dim view of the price of packing-station eggs.

We also had an enthralling time on the subject of how many rooms each house contained. I was supposed to find this out in my official capacity and the book of words with which I was issued expected that there might be some difficulty in getting this information. Whoever wrote it bent over backwards in stressing tact and politeness. But I needed no diplomacy to get an answer. Every housewife burned with feeling in the matter and, what was more, she yearned to give her feeling words.

More Room

They may all be for small houses in the town, but in the country our desires are quite contrary. We want more rooms, closets, cupboards and porches. Bigger ones, too. This is not to be wondered at when you think how much room country clothes take up. Six pairs of rubber boots can crowd out any normal cloakroom, and when you add six large coats to the pegs above the boots the room bursts at the seams.

I do wish some architect before bending his mighty mind to planning the countryman's house would do a simple sum in the number of clothes the country dweller needs. He must have two of everything; one set to dry the other. Besides the number of garments it is wise to consider the kinds of materials. The kind of waterproof you wear to go to the steading in is not the natty slight affair you don for the town, but a good strong tweed affair as like a sheep's fleece as you can find. I think the real answer to rooms in the country is to build four above the number you first thought of; even then you'll have to dry one overcoat on the trailer radiator.

THE COSMOPOLITANS

27 April 1948

Like every other farm in the world, we are dependent at rush seasons on casual labour. Spring with its tattie planters, summer with its hoers, autumn with its harvesters all see the farm with an extra squad on.

Farmers who live near a town or in the centre of a busy agricultural district can rely on finding such labour fairly easily, but folk like us who are marginal land people have to depend on the labour officers of the Executive Committee to find us the squads we need.

This they do very efficiently, and the only thing we grumble about is the fact that this labour can't speak English, far less the doric we use among ourselves.

First there were the Italian P.O.W.s. I well remember the misgiving we felt the first time we were confronted with a batch of men who did not understand a single word we said.

Yet with good will on both sides we soon managed to evolve a pidgin doric which sounded awful but served its turn.

With all modesty, I may say that by the end of our first harvest together we were speaking a fluent ungrammatical Italian.

But alas, before we became really proficient, we lost our Italians and got landed with Poles.

Dashed but dauntless, we discovered that school French was useful here. The silly so-and-so's could not understand our French pronounced with a good doric accent, but we could always write.

You've no idea how slick we got, and were almost at the stage of shrugging our shoulders when the Higher Ups said "No more Poles. Here's Germans for you, just a wee changie."

45

I shan't conceal from you that by this time I was feeling pale. Others of my friends were livid. You will gather we were all upset.

The German turned out, however, to be serious minded and had a grammarian's interest in language per se. Soon we shot our Romance languages overboard and hung in with determination to basic Anglo-Saxon.

We found it quite an experience diving through seas of nouns, adjectives, qualifying clauses, to come up for air as it were with the ultimate triumphant verb in our mouths.

Yes, I may say again with modesty (I hope) that we were coming on a great with our German.

Then to-day I phoned the office to say I could do with a squad to get the rest of the tatties in. "Oh, but yes, surely, Mrs Macpherson. We'll send you Yugoslavs."

Cosmopolitans, that's us. Ach, you've no idea.

PROPER USE OF LEISURE

When the bonny days came and we were able to turn all the cattle out from their winter pent byres and folds, we were faced with the problem of what to do with Frankie, our handsome-in-a-flashy-way Ayrshire bull.

Obviously we couldn't send him to the hill with the heifers, so we tried to keep him with the sedative cows. This would have been all right, but the cows spend a lot of blameless time cropping about the house.

The grieve and I, who are on quite amiable terms with Frankie, didn't find it the least unnerving to meet him horns on in unexpected places, but, unfortunately, visitors did not see it in our light and with reason I must admit.

The day the tattie planters arrived Frankie uttered a shrill scream of fury and made at their car from which the petrified men bawled at me; - "Hey, wifie, come and tak' awa' yer bull."

Very courageously, with the collie at my heels, I headed his taurine majesty into the nearest gateworthy field.

After this there was only one thing to do if we weren't to immure him in his winter quarters, which we disliked doing not only for his sake but for our own, for we had no wish to turn him sulky.

That was to cut off a bit of field and fence it in high and strong. This we did, and Frankie wasn't nearly as grateful for it as he should have been. Cos why? Because he was bored.

Nature, he inferred, was all right, but what about a spot of company. Not for the first time I wished he could read, and then I could have sent him "The Bulletin" to keep him out of longing.

My sympathetic heart was wrung, and goodness knows what maudlin stupidity my sentimentality might have led me to if Frankie had not suddenly discovered the solution of his own problem.

Left behind in his enclosure was a wooden kiln used for hearting stacks. One evening as summer veiled the hills with sweet twilight Frankie found it. He set to partners, stamped, tossed his great head, and finally gored it.

At last all was over, the sham battle done, and he raised his victorious head to listen for the plaudits of an imaginary audience.

There was such an air of florid élan and bravura about him I could only open the window and shout "Bravo" Frankie looked and accorded me a bow.

Since then we have a performance every night. Frankie is no longer bored. Neither am I.

TIME MARCHES ON

3 May 1949

As April grows into May, the hill farmer makes a last desperate effort to get his ultimate crop in. Up here the growing season is so short we must try to take advantage of every day of it, so that we must work all hours to get our seeds in.

No matter how well on we are with our cultivations, we still have but a bare month to get our corn and silage in, our tatties planted, and our most necessary manure sown.

If April were not so capricious it would help, but what can you do with a month that gives you two days of real summer heat and follows them up with a snowstorm lasting as long?

Ach well, we can but wrastle on feeling as usual that we are performing an act of utmost faith in putting any work at all on to the unresponsive earth.

Last week we got home 25 tons of lime, and the men spelled each other at the tractor to get it all on to the sown-out crop. It blew and blew all the time they worked, so that the fields looked as if there were a localised snowstorm raging, and at the end of the day men and machines were so encased in dust that they looked like robot characters out of an old-fashioned German film.

I could not resist pointing out to the grieve that under the circumstances it was a blessing we were using mild ground limestone and not the vicious love of his heart, burnt lime, which really does burn.

Of course, you need only half the burnt lime that you do of the ground limestone, but it is so brutal to handle that I do not urge the manure merchant unduly when he says he can't supply it.

The incessant wind has been a nuisance, too, in other ways. It has held up our grass seed sowing. Grass seed is so light that even a puff of wind can send it over the hills and far away, leaving your field bare.

We have had to put off sowing it several times, but at long last it is in and I breathe another sigh of relief at another difficulty overcome.

Yet for all spring's laggardness the tide of green rises irresistibly up the valley.

We got the heifers into the grass and they nearly swooned with delight at the change from woody turnips and chaff messes, which were their monotonous diet for the past month.

The humans, too, were mightily relieved to be rid of the everlasting mucking about with treacly water and oat dust which the cows had for their morning cup of tea and without which their milk supply would have vanished.

Grass is a very good thing and thank goodness you don't have to wonder if the beasts are getting a correctly balanced diet. Grass is their perfect food.

But as well as the grass being up the oats are through, too. Millions of their green, gallant spears are piercing the earth at this very minute.

I could weep tears of silly joy at the perennial victory of spring.

THE LONG HAUL TOWARDS SUMMER

1 April 1961

That the official opening of summer should have been marked with a gale and wild squalls of snow was to be expected, for Scottish weather, especially Scottish hill weather, is a specialist in wersh humour. So we cowered by the fire as much as circumstances would let us, and longed for the cessation of the blast. After all, we have had a wonderfully safe passage through winter, and must feel grateful. Yet however mild winter may be there is no gainsaying the arduousness of the long haul from Christmas to Easter when there is no light, no growth, or even any of the small companionable noises which ameliorate isolated country living.

By the beginning of March one has developed such a craving for signs of authentic spring that the smallest tassel on the larch, the merest thread of birdsong can make the day for you.

It is not only because we collect every vestige of spring so avidly that we are so cast down by the abrupt storms which mark her erratic passage. A sudden violent change in temperature sends us into a paroxysm of anxiety about our beasts, and we recollect with painful clarity other occasions which left us with pneumonias and agues and deficiency diseases.

We are not so mawkish as to attribute to the beasts of the field emotions and feelings outwith their make-up, but we cannot escape feeling that they and we share a common mortality. Because of them we dimly glimpse what the saint meant when he spoke of the "whole creation which groaneth and travaileth together." For the brief space of their lives, too, the beasts depend on us for care and succour.

That is why when the wind blows and the air is thickened with snow we struggle into coats and big boots to scour the inhospitable world for animals which may not have found shelter. To see all the cows under a roof and their calves warm beside them while the tempest rages without is a wonderfully comforting feeling.

New Lamb

And storms can have their triumphs too. At the height of a gale out of the east, Dod came in with a lamb a bare day old. The small head lolled over his arm and the tiny body lay flaccid. We stoked the fire and put the nearly dead creature to warm by it. For more than an hour it lay inanimate and then slowly the minute life began to creep back. We cherished the little thing before the fire and when it uttered a faint cry we were nearly afraid to hope. I went to rummage out an old teat used for this purpose before, and I filled a bottle with warm milk and glucose. But the cold little tongue refused to function and we really thought the lamb was dead, when all of a sudden it began to suck and we knew it would live. We tidied away the squalor inseparable from mortality, but all we knew was triumph. The lamb slept by the hearth.

RED, WHITE AND R.M.S.'s

R.M.S.'s are the forms you must fill in before you are allowed to buy petrol. R.M.S. Six, imperially printed in purple (and kindly supplied to me by a sympathetic reader who has thus saved me many a 10 mile journey) is the one you fill in for pink petrol for your farm tractors. R.M.S. One is the shy little fellow with the jaundiced face that you fill in when you want petrol for your car.

If I was a great, big, enormous, successful farmer I should not have a car at all. I'd telephone the garage to send me the latest thing in vans or lorries or jeeps, and then I'd get lashings of pink petrol with which I'd swoop and leap and zoom from sale to sale, from show to show, nae bother.

But being a she-crofter I can't afford to buy a commercial vehicle, but must make do with my 1938 saloon car. So I must fill in R.M.S. One to get petrol for a car which has to do all the things a commercial vehicle does.

Mind you, I find the Reg. Pet. Office a most understanding place. Provided I can advance a sensible claim for petrol I always get it. But not all my farsightedness nor the Pet. Office's sympathy can entirely cope with the things that happen on a farm.

I mean, who could foresee that the last couple of calves we bought should have been despatched just in time to miss the last connection up to the benighted regions which we inhabit?

The result was that a pair of indignant and highly expensive youngsters arrived at Forres station, ten miles away at eight pip emma of a cold wet night.

A thoughtful young man, urged, no doubt by the unfuriated bawls of the calves, rang up to tell us what had happened. There was no further train that night, and if we did not come for the calves they must remain in the station all night.

Now perhaps we ought to have taken the tractor and a cart and made the long, slow journey down the country. Being human, with sympathy for the calves as well as for our own bones, already weary with a long day, we took the back seat out of the car and went for our livestock in half the time the tractor would have taken. Of course, it cost us a whole gallon of the precious white petrol.

I snort with fury. Can you hear me?

NON-PRIVATE LIVES

29 April 1961

The census has been the occasion for many people to make impassioned pleas for the necessity for respecting personal privacy. Theoretically I am all for their point of view. Practically I do not see how it could apply in the case of the farming community. As a class, of course, we are chock-full of suspicion and as anxious for concealment as the next; but the nature of our calling compels us to pursue it in the broad light of common day and on a scale which defies obscurity. Whoever heard of anyone hiding a 20-acre field corn? Not even a wee half-acre of tatties can be consigned to the skeleton's cupboard. What we do is there on a landscape for all to see, admire or disparage.

A friend put it very well when he was discussing with Dod the latter's decision to reseed grass directly instead of using a nurse crop which is the practice here. "Well, George, I'm some feart. But it will be *interesting* to see." The number of his rolled "r's" was an indication of his satisfaction that his curiosity could not fail to be satisfied. Considering the park in question borders the public road, I am sure that not only Willie's curiosity will be satisfied but every other neighbour's as well.

If you cannot veil your field work in decent obscurity still less can you gloss over your policy. Ayrshire cows in a pasture mean that you are dairying, while black cattle spell out beef. Blackfaced sheep on the hill mean a pastoral farm with difficult cultivation on brae set fields, very different from low flat acres, which grow cereals on the grand scale.

Registering

Perhaps the folk who are slickest at reading the private lives of the farming community are the Department officials. As they drive along their minds automatically register not only how many acres lie before them but they can tell to a month how long that particular field has been in grass. They are as likely to know how many calves you have as yourself, though I should not say they are any better than your neighbour in picking out which cow was bred on the place and which cow was bought in. They are probably more knowledgeable about your hill drains than your neighbour, but I doubt if they have anything on him with regard to the amount of heather you burned.

I have sometimes toyed with the idea of keeping a new implement dark, but here again I am thwarted. Naturally I know you cannot hide a Dutch barn, but I did consider that a small thing like a Cambridge roller could lurk secretly in the Nissen hut. Well it cannot, for every ridged field where it was used advertises its presence. For the same reason everyone knows when you get new discs, and, of course, a new tractor, all loud and glittering in fresh paint from the back of the lorry which delivers it, announces its arrival to the whole community. Large machinery like combine harvesters and hay balers are impossible to ignore for they flaunt themselves with horrid notoriety in open doors of cavernous sheds about the farm square.

We cannot even keep our money affairs private. There are our investments walking about on four legs all over the place for any country Tom, Dick or Harry to count. So many black cows on a marginal farm mean so many pounds in hill cattle subsidy, so many black calves disporting themselves on the moor mean so much on the stots, so much on the heifers, after the Government puncher has marked them. Privacy? We just do not know the meaning of it.

TELEPHONING ABOUT TATTIES

4 May 1948

It was a wet day, so we spent it redding up our various bags of ware tatties – which means potatoes for you to eat.

We are not primarily ware growers, but deal instead in the seed market. Nevertheless, in spite of all precautions in spacing and so on, we find that after dressing out our seed we are in possession of a sizeable quantity of ware potatoes. These we sell through our usual merchant.

So we duly weighed, labelled, and stacked our eating tatties in readiness for loading, and then at the back of the shed we came on a ton of Edwards. I had the bright idea of phoning our merchant and getting him to accept these along with the others, since it would mean an obvious economy in transport. All this happened, you must remember, before rationing stopped.

But the merchant nearly passed out at my suggestion. If he as much as laid a finger on an Edward he'd be flung into durance vile and branded as a felon. The reason? The Ministry of Food said "No moving Edwards or else –."

So I rang the MoF. My, you could hear the shudder with which they greeted my unorthodox approach. Wot, no forms?

So, said I, what way could I have forms when I didn't know I was going to have the tatties, owing to me being a poor miserable crofter that grew only seed tatties, ware tatties being just an embarrassment to the likes of me?

But as there was a scarcity of potatoes (or was there?) maybe the MoF would like to do something about them.

So they said, very informatively, "Edwards are the best keeping

57

variety we have, and we must keep them until the new potatoes are in'. Then I asked when that would be. So they said 'There is no demand here for Edwards."

Amiably I agreed that I thought they were just about as tasty as soap, but I still wanted to sell them.

Fancy asking a great Big Important Office to buy one miserable ton of tatties? It's time the crofters learned their place. But still they did not lose their temper, but just said 'Why not just plant them and don't bother selling them?

Admirable solution. Such a pity though that in spite of subsidies farmers still have to market their produce. Wouldn't it spare the "offeecials" a lot of inconvenience?

So the merchant couldn't, and the MoF wouldn't. Hey ho!

Mind you, I'm lucky. A lot of my neighbours haven't got their Edwards bagged in the steading, but lying in pits in their fields.

By the time they're allowed to touch them they shan't be able to get near them for the waist-high corn. They'll quite likely be loony by that time, so it won't matter.

Exit in hollow laughter.

NOW IS THE MONTH OF MAY

6 May 1961

Nothing more clearly differentiates the Scots from the English – or for that matter from the rest of the Europeans – than our attitude to what is for everyone else in the northern hemisphere the traditional opening of summer; but which for us in our rugged latitude is one exasperating storm after another. Not for us a May Day parade with banks of flowers and hero cosmonauts; gaunt and girning we survive a "lambing storm" and gutter about with reviving lambs staggering about the stone flags of our kitchens while their half frozen brothers come to in hay boxes before the fire.

Incredible
"Lenten is come with love in toune" they sing before the dawn of English poetry and we suppose it is true enough, but while the wind scourges the landscape and hail sows the nearer hills we find it a trifle incredible; anyway, who has time to find out when there are all these lambs to inject for revolting diseases like 'pulpy kidney'?

We have hardly hung the crook on the wall and the lambing balsam in the vet cupboard when the next of our traditional storms breaks. Up out of the east comes the "teuchat storm," and with it the peesweeps skirling like angry kittens at the rainbows flushing and fading between squalls. Mated grouse cackle their remonstrance on the hill where the harsh deer grass makes its courageous green splurge against the dun heather. Storm cocks, these melodious Cassandras of bad weather, hop from beech to larch saying "I told you so yesterday." These thrushes are some of the best weather prophets we have and it is sad to hear their clear triumphant song on a fresh, calm morning and know that all that pure joyous sound is but the herald of next day's wind and rain.

Drying Up

Farmers anxious to sow barley go round in a tension of impatience but at least they do realise that at this time of year the ground even in Scotland does dry up quite quickly. They will therefore manage to get the crop in just before the "gab of May" puts in its unwelcome appearance.

In England they have maypoles and morris dances. We are a bit hazy about the maypole but the chap who wrote "the nine men's morris is filled up with mud" is the bloke for our money. And when we come to think of it perhaps, too, the English knew the maypole idea was a rash one. Look at that Jassie who was all for being Queen of the May one year and Alfred Tennyson found her a twelve-month later dying from the cold she got then. Why she put flowers in her hair and forgot her rubber boots we can't think.

We are sorry for the lass and consider her fate as a warning, as indeed we ought, for we have yet to come through 'Buchan's cold spell'.

We have now used up all the descriptive names for May storms, you will notice, and must fall back on more mundane ones. The cold spells sees the rest of the month's thirty-one days past and, thoroughly weather beaten, we lurch into June.

And yet all Mays are not so uncomfortable as I make out. Today as I write the cuckoo calls in the hundred hidden ravines which line the banks of the river. Sun shines; grass grows; and it would ill become me not to return thanks for all the gracious benefits thus bestowed on me and all the rest of the glen. We do after all live in Morayshire, that county which the old monks described as existing "at the back of the north wind" where there are forty more sunny days than anywhere else in Scotland. Today it is easy to believe it and only human to hope that it will last.

END OF TWO WARS

Even if the weather had been good, I don't think we should have done very much on the farm this week. As it is, May rose on the wrong side, so that all we've been able to do has been odd jobs which don't take us very far from the wireless.

It has been a strange feeling, this waiting in the wings to watch dust and triumph mingle. Odd, incongruous thoughts chequer the mind.

In one instant we think of returning prisoners of war who once were our neighbours in this countryside and who have lost their young years since. In the next breath we think of how we must change the rotation of our fields which war-time cropping has sent all awry. Marginal land will not support this cavalier treatment indefinitely, and we must get it back to something more normal.

Suddenly we become aware of our own personal good fortune. Almost alone amongst our fellows we have not been called upon either to manufacture or to man engines of death during these bitter years. The work has been unrelenting and incredibly hard, but it has been concerned with the normal things of life and not with the grotesques of death.

I cannot help thinking just now of Armistice Day in 1918. I was a child then. We lived in a small fishing village by the Moray coast, and on that still, November day half a lifetime ago there were no crowds, no gaiety. There was nothing but quietness.

Peaceful Horizon
By some chance I was alone in the echoing house and childlike I ran to look for friends. Outside on the flat coping-stones of the sandstone

dyke bright lichens grew, patterning the stone like a lizard's back in orange, scarlet and green.

There seemed no one in the wide calm world. I sat swinging my legs over the dyke, and I thought – "I'll remember to-day always." Yet to a child there was nothing to remember. There were no crowds, no bunting, no cheering; just nothing and a peaceful horizon.

THE CREEPING MIST

11 May 1948

The long spell of dry weather in early spring, while it enabled us to get our crops in with the minimum of fuss, threatened to undo all its good by overstaying its welcome.

However, we need not have feared, for the end of April and the beginning of May have brought us all the rain we needed. First we had snow and then hail, followed by frost and then rain, and rain again.

To make our discomfort doubly sure we then endured a "seapiner" which is a three-day spell of rain and mist accompanied by a faint but persistent north wind. No wonder I'm aching with rheumatics, and two yearling heifers are down with severe chills.

How I do dread these storms of rain, preceded by flights of sea birds forever hurrying inland from their sad homes by the shore. Banks of mist sweep up after their swift wings, and soon the plain is quite hidden from us, while the hills stand round suffocated under the trailing vapours.

Underfoot there is mud everywhere, and since our cows are out on good new grass only another farmer can guess the state of our court.

I shed my boots and wet coats and dungarees in the porch, misliking myself much in the permeating odour of byre.

Wet cloths, wet floors, smoking fires, everything is here for tears. And yet for all the vanished beauty of the day, all is not lost, for when grey twilight follows the grey day the familiar sodden landscape assumes in that half light significance and dignity.

Mist coiling up from the coast ceases to wear the air of that rococo ballad of Kingsley's Mary who went to call the cattle home and instead becomes the attribute of an apocalyptic world.

The great elm and ash which guard the road on the other farm magnify themselves and, ceasing to be trees, partake of the life of Ygdrasil and become at once the gates and guardians of a hidden land. Slowly the moving mist changes the whole landscape – now obliterating half its features, now adding stature to others.

The great fir woods are erased, but the junipers and whin are suddenly huge with secrecy.

Only perfume lives where the birches were, and the full-running river is invisible. The steading is lost, but boulders in the fields' verge are keeps and castles, while the ordinary gate leading to the grass is no longer a gate but a parallelogram in infinity.

Incessant flights of birds complain in the wet obscurity. Owls shriek in the thicket, and it is night.

Sighing, we pray for sun tomorrow.

AT THE TERM WHITSUNDAY

13 May 1961

"The entry shall be at the term Whitsunday," are the words which on most Scottish farm leases indicate to the incoming tenant when his occupancy of the new farm begins. Whitsunday is our Scots quarter-day and as it falls on May 15, it is not inappropriate to-day to consider the implications which, on Monday, so deeply affect the business of the countryside.

For many years it was nearly impossible to rent a farm since tenants who were sitting pretty as regards security, rent, and farm prices were understandably determined to stay put. But even in rural Scotland economics (chivvied by exasperated landlords) catch up with life.

Therefore very cautiously a few farms began to creep on to the tenant market. Rents also jumped dramatically as demand exceeded supply. When the lairds saw that not only were rents doubling and even quadrupling, while farmers were willing to take a realistic attitude to farm improvements, they started to detach some of the places they were holding in their own hands and set them up for tenancy. The general rule of course is still for the bigger holding to be considered the more economic unit.

New Tenant

The knowledge that a farm is to let throws the whole countryside into a steer of conjecture and interest. Russians and Americans can career as they like in space, but we, since we are by the nature of our calling more earth bound than most men, think the project no more than an interesting irrelevancy and not nearly so close to our hearts and bosoms as the policy the new tenant in the Bog or the Muir or the Hill must pursue in order to pay the £800 they say he is giving as from this entry at the term of Whitsunday 1961.

Avidly we read advertisements about the displenish sales where outgoing tenants sell off their stock and implements; and we eagerly make arrangements to attend said sales. Since by this time all our rush work is over we feel we are free in our conscience to take time off to go to the roup. Anyway there is a strange moral compulsion on farmers to be present at other farmers' sales. It would take a better psychologist than me to unravel the passions and motives that move us on these occasions. Of course we want to see another man's business, but not just for curiosity. We like to know what he has accomplished with the means at his command, we can admire his achievement and sympathise with his difficulties.

Big Day

We also have come to "give him a hand" as we say. That is why we bid far more highly for his beasts today than if he were selling them in the auction ring at the mart. I am not saying that we are so impeccable in our motives that we are above a bargain among the hand tools, but we really do not forget that this is our neighbour's big day; this is one of his few great occasions. By our bidding we express our hopes for the comfort of his retirement and by our cheerful greetings to him and to our fellows we confess that we are all bound together in the bundle of life. Perhaps, too, we are realising more closely than other men how indissolubly we are bound to our great mother the earth; she is our alpha, our omega.

The crowd jostles and flows round the steading through the refreshment tent. The confused cattle bawl their unease at the strangers while trucks and floats lurch over the parks to collect purchases for new owners. For a brief hour the farm lies patient and unpossessed till Monday "at the term Whitsunday."

TEEM TRAIVEL

18 May 1948

This year for the first time in our lives we upland farmers are no longer going to pay income tax on three times our rental (it being under £100), but are going to pay on our income.

To all of us the arrangement seems equitable, but it has a much deeper significance than a mere monetary adjustment. For the first time we must now keep books. What a world of revolution lies in these simple words!

Most of the small farms hereabouts are family farms. If books are going to mean anything at all, they'll mean that each member of the family is entitled to the minimum wage. So that a farmer with two grown sons working for him is going to have to find £9 every week less a board and lodging allowance.

It is a lot of hard cash to take out of these frugal acres.

Well, shall the farmer let one son go? If he does, he's up against a flaming difficulty. There is so much work to be done on these marginal farms for which there is no immediate profit. Take picking stones off our granitic fields for instance.

The saving on your implements is immense, the subsequent cultivations are easier, and the crops benefit immeasurably from having the weeds really dealt with instead of having them lurking behind boulders ready to muscle in the moment a cereal crop is down.

But paying someone £4 10s a week to pick stones off your field strikes me as what the grieve calls "teem traivel."

When he is working in an ill-shapen field where there are lots of gussets he does not get the full good of his implements and must "traivel" quite a long way with them empty or teem before he can get them into the ground again.

There's lots of "teem traivel" on our upland farms, and keeping books is going to make us face up to it. We shall find it irksome and unpleasant.

What, for instance, are we going to do about the tractor men we engage for the rush periods in spring and autumn?

Nobody here dreams of charging these men board and lodging. Yet food and a roof cost money. And we can't put it down as an expense, because these men are given an allowance for their board.

It is quite a dilemma for a hospitable but careful people.

Yes, we are going to have to look hard at our "teem traivel." We're going to lose a lot of graces and kindliness, but we're going to be efficient.

I expect we'll find it quite pleasant once we get there, for there is a lot of comfort in efficiency, and it is surprising how often "teem traivel" is just another name for mental laziness.

WHERE "THE BROOM BLOOMS BONNY"

27 May 1961

All day long you can hear the scrape and clink of the roadmens' shovels as they work on the stretch of highway between here and the little bridge behind the birches. They are busy tidying our moorland verges and hacking down the broom which bursts in great golden fountains from old stone dykes to make an already difficult road less visible than ever. At midday we see the blue smoke of their fire rising above the small green leaves of the birches, and cannot help envying them the surrounding beauty of their picnic.

Thin Sand

At this time of year whin and broom blaze across miles of upland waste in extravagant splendour, gloriously at variance with the thin sand which nourishes them. It would be a withered heart that did not feel itself uplifted at the sight of such casual profuse beauty. I like to think it is not only the fact that oats are still wallowing in a financial nadir that keeps us from reclaiming the scrub acres – even at a subsidy of round about £12 an acre. And we can comfort ourselves, if we have to, by reflecting that the close-grown thickets of both shrubs afford much needed shelter to our out-wintered beasts who have to endure the rigours of the bleak moor.

Before the war we were on this farm sheltered most kindly from the weather and the Dava by woods which grew on three sides. When the wood was cut down we spent a chilly interim in the blast till the broom moved in. Nature is kind even in the stoniest places, and she abhors scars as much as the proverbial vacuum. This year we burned rather more hill than we had calculated on, but when we went to look at it the other night the tough hill grass was establishing itself there along with the minute plant of the needle whin all decked out in pure new gold.

So the broom took over here when the trees were cut, and it has been with us ever since, though it has had to give up some of its further colonies. The ravaged forest has been replanted and saplings must have room to grow. Hundreds of foresters with hundreds of hooks cleared out the broom but it bears no ill will. It crowds right up to the wire netting which protects young plantings from vermin and gaily beckons the puritanically marching ranks of earnest conifers to a more cheerful and relaxed life.

If the conifers are deaf to such temptation the birds are not. The other morning when I went out to feed the hens there were hosts of yellow hammers breakfasting on the leavings from the night before. They appeared to think their habitat so appropriate that I was the intruder not they. When they were quite ready, they moved off into the blossomy thicket and perched heavily anywhere. Drugged with the hot velvet smell which even in early morning hangs on the summer air, they gazed at me sleepily before uttering their little reedy song.

Run of Notes

They can sit for hours thus, golden birds upon golden boughs shaking out the same small monotonous run of notes followed by one or two longer cadences. The English say the yellow hammer is saying "A little bit of bread and no-o chee-se," and the Scots say he is reciting "Deil, Deil, Deil, ta-ak ye." Our folk lore tells us that the devil gives the "yellow yite" a drop of his own satanic blood on a May morning to mark his eggs. We are a wonderfully familiar lot of folk with the devil. I am sure there is a moral for us somewhere but, like the yellow hammers bemused with the heavy sweetness of the broom, I feel too happily soporific to work it out.

SUMMER

LUNATIC LANDSCAPE

2 June 1968

L ast week but one we were basking in a heat wave that charmed the seeds through the earth, set the cows free of the byre at nights, and made us run to abandon the chrysalis of our winter woollies.

And now what are we doing? This morning I walked to the byre over sheets of ice, moulded into mail by the incessant snow showers of the past few days.

We have hurriedly resumed all our heavy clothes again, and the cows not only spend the night under a roof, but also parts of the day as well.

Each wildered morn the astonished sun rises on a world that is one large lunatic landscape. Low hills entirely snow clad are at strange odds with acres of yellow-flowering whin and broom. "The fruit-like perfume of the golden furze" hangs unnaturally on the wintry spears of the chill air.

Blackbirds flute in amazement amongst the apple blossom boughs miraculously pink and white in a garden where verges and fences are deep in snow. The grass grows thick and green about the greyly frozen pond, and new clover shudders amongst its sheltering purple-brown rye grass.

But we could stand the curiosity of a landscape that looks like Paul Nash if we could escape its inconvenient implications.

A prolonged spell of cold weather at this time of the year is bitterly trying. Alas for the early neep breer which was no sooner born than blasted.

We make a gloomy note to order fresh seed to resow our ravaged acres. The cows who ought, of course, by June to be standing knee deep in nice, nourishing grass are pretty well byre bound.

Our winter feed is done, so shall we put out the cows for a feed and let them also catch pneumonia? Or shall we keep them in their stalls and let them starve? Or, like the grasshopper, shall we let next year's calf cubes already stored in the barn go past the hypothetical calves and squander them on the cows now?

Evening with its long light comes to mock the ice in pools and buckets. For an hour or so the world is calm, untroubled by squalls of hail and sleet. The sun shines and, far to the west, Wyvis stands snowclad and shining like the bastions of God. All its shadowed valleys are doors to glory.

Oh me, shall we put the cows out or no?

HEIFERS TO THE HILL

8 June 1948

What with the drought in early spring and a blizzard a few weeks ago, the hill pasture was looking fair drumuid.

As a result then, we were not able to take in our stranger summering beasts as early as we and their owner would have liked, nor could we move our own beasts from the home parks, which by this time were sorely needing a rest.

However, the weather clerk in a moment of aberration allowed two bright and balmy days to escape the sack of Aeolus and the herbs and grasses on the hill sang up green and fresh from the moss.

Without more ado we sent word to the man who'd hired our rough grazing and held on our own behalf our annual summer round up. That is to say we divvied out the animals who are to remain at home and those who are to be sent to the hill.

When you rear attested dairy beasts you must watch that during the second year of their lives they do not grow either too coarse in the bone or too broad in the beam. If they do the milk vein will become constricted and, naturally, in dairy beasts this is a major fault.

Summering on the hill guards against this. But you must be careful not to send them to the rough too early, for if you do you will discover that the spartan faring will stunt them instead of hardening them.

As a general rule then what we do is to send out everything over a year old and keep at home the younger beasts.

But in farming as in everything else that has to deal with living creatures, blueprints do not always work. They certainly don't work in the case of the big Fanny-Annie.

She is a swinging big Ayrshire calf about nine months old. Last year

75

she was suckled along with another. But while small sister allowed herself to be weaned with pleasant co-operations big Fanny made a most infernal din about it.

We kept her in the fold for a week, and she bawled the farm down. We put her out in the nearer field and she cleared everything, gates, fences, and any other obstruction with angry ease till she got herself back beside her foster-mother, Venezia, who by this time was dry and in no fit state to be bothered with a galumping daughter nearly as big as herself.

Hurriedly we sent her to the other farm, where she behaved herself well enough till a freshly calved cow with three calves at foot was sent down also. Thereupon Fanny-Annie elbowing out the rightful occupants, proceeded to help herself to their meal ticket.

There's no help for it then. Against all rules and proper practices, there is only one place for the brute and that is the hill. And there she has been sent.

BIRD COMMENTARIES OVERHEARD

10 June 1961

Not the least pleasure in country living is listening to the accompaniment provided by the non-human creatures who share the world with you and make their own observations on its joys and hazards. If you want to know how inadequate human speech can be, listen to the thrush, obsessed with the morning's loveliness singing in the green choir in the beech. The lark praising the risen sun must ever have the ultimate word, while even the sparrows plebeian and chattering angrily on the roof at a sudden thundry splash, can express irritation better than you, even if you can hold a pen and look up the dictionary.

Spring and summer are the great seasons for birds and their conversations. You can hardly put your head out of doors without hearing a running commentary on the universe. The starling anxious, for all his fancy shot-silk waistcoat, is never finished remarking on the insatiable appetite of his family, while the peewit is worn down clapping his hands to frighten away the gulls who might menace his babies running about on stilts among the marigolds at the foot of the burnside park. The cuckoo mocks the noon from a hidden wood, and the curlew woos interminable solitude on the high moor.

All day there is a soft continuous noise, always beguiling, always companionable; but at some hours it is more noticeable than at others. At midday, for instance, there is no more than a sleepy murmur, while at four in the morning the blackbirds are whooping it up with a trumpet voluntary.

Silver Rings
Swallows need an early morning cup of tea before they utter, and,

anyway, they let themselves be diverted from the subject in hand by any passing cloud of gnats. Swallows are always breaking off to tear off after some insect, and I lose the thread of their discourse when they interrupt it to rip off great breadths of blue silk air. Beautiful and graceful as they are, I do not find them nearly so restful as the willow warbler. On a hot afternoon nothing is so agreeable as to listen to him spinning silver rings across the lazy sunshine, promising soothing things like afternoon tea on an Edwardian lawn.

After 12 hours of comparative silence the blackbirds are in voice again and, come the evening star, they are ready with their serenade. The goat-sucker, who has been dosing away all day, pretending he is a moth and not a bird, now thinks the hour has come for him to read the 10 o'clock news, which he broadcasts for a full half-mile on a spinning, reeling, wave-length. In our long Scottish summer dusks he darts and turns for hour on hour, and, as he flies, he croons his curious song, at once far away and near.

Cock Pheasant

After summer goes the birds fall silent, but, because they do so gradually, you do not notice. Then you are hanging out the washing or walking up the hill road to open a gate and you realise with a pang how alone you are. The grouse calls harshly from the moor and the cock pheasant, flighting heavily and gorgeously in the thick planting, utters his single raucous cry. From now on the only accompaniment to your country day will be the abrupt ejaculations of autumnal birds facing winter.

And when winter comes the robin will leave the pearl necklace of his song on the garden fence and owlets will gibber in the moonlight as they smell an approaching storm. The fox will bark back at them and the roes will cough. In the high tops the red deer will roar their defiance at one another and crofters in the valley will hear the noise echoing and reverberating in the hollows scooped and drifted by the corniced mountains.

BRINGING IN THE TOURISTS

17 June 1961

Sales of work, like cattle shows and marquee dances, are an essential part of the summer scene in the country. Unlike the dances and the shows, their purpose is strictly utilitarian, and any amusement value arises strictly on the side and on the part of the observers.

To run a sale of work successfully needs equal parts of ingenuity, tact, and hard graft. These excellent virtues can be wonderfully reinforced by the goodness of the cause for which the sale is being run, and since country sales of work are practically church monopolies, the justice of the undertaking is self-evident.

The first great difficulty to be tackled is the date. Now we live in a tourist-conscious area and one of the things we expect of the holidaymaker is her appearance and purchasing power at our sale. This insistence on the tourist schedule severely limits our choice of date.

In spite of adjurations the holiday season remains obstinately stuck in the three summer months, June, July and August. Three months, 12 weeks, looks elastic enough, but, because every other neighbouring rural district has the same idea as ourselves picking a favourable moment involves more craft and even more strategy than you imagine.

However, having got ourselves a place in the queue we feel we can now turn and deal as best we can with the domestic exigencies of the farming year. Supposing our spouses decide to start hay-making on our Big Day? Perish the thought since we need not only their brawny arms to put up the trestle tables we use for stalls but we also need the family car to take us and our contributions to the venue.

After all this jockeying with dates I look back with longing on the pre-summer visitor days of my youth, when the fishing village where we lived never knew the name of tourists, since the only summer visitants we ever had were the nuns from Elgin who passed decorous days reading holy books by salt sea pools, and we took them as natural parts of the scenery. We didn't have to bother with cars either. The stall-holders piled everything into their creels and up the kirk brae with them, and that was that.

Of course, basically, sales of work are always the same. Work stalls, cake and candy, white elephant and tea stalls. I never look round a country sale of work with its enamelled pails full of peonies and pyrethrums, and its anxious stall-holders wearing frocks to match the flowers without thinking admiringly of all the unwearying endeavour that has gone to arranging the business.

You can smell the hot kitchen where the feather-light sponges were made, you hear the whirr of the sewing machine running up the aprons for the work stall, and you doubt if your own reserves of diplomacy would have lasted out half as long as have those of the different conveners.

After 50 years of sale of work attendance I am no nearer solving the puzzle of the tea stall. The tea is the nub of the country sale of work. Everyone expects, and gets, a smashing meal, but the charge is always a paltry one and sixpence. The "teas" have always the lowest takings. But perhaps the indomitable women who run it remember Phebe and all the other women who get an honourable mention in the sixteenth chapter of Romans. It is comforting to know that the early Christians had to do a lot of "scuddly" work too.

THE "HIGHLAND"

22 June 1948

By the time this appears in print I hope to be at the Highland show, which this year takes place almost next door in Inverness, a mere thirty to forty miles away.

Since the war the "Highland" has been in abeyance and the farming community has been deprived not only of much scientific knowledge in the way of implements and the rearing of livestock but of the fun and friendship which coloured the whole year between one Show and its successor.

For years the "Highland" has been the rallying place for Scots farmers, and we looked forward to the event as much for social reasons as others. No wonder then that for weeks we have all been scheming how to get ourselves there.

First there is the little matter of conveyance. In a glen 10 miles from a main bus route it's not very easy. The trains are miles away and show an aristocratic indifference to the conveniences of commonfolk. Lots of us have what pass for cars, but lots of us haven't.

For myself I'd like fine to go on the tractor. It would be fun to make a triumphant entry into Inverness with its muddied, cobbled, Hieland streets seated on a bellowing Fordson. Besides it would save on the white petrol.

However, seeing a number of sedate neighbours, including the grieve, are going by car, I'd better dismiss these rakish longings.

But having settled our conveyance, we have also to settle the work. You can't just walk out of a farm.

When I broached the matter to the grieve he said (of course) it was out of the question – (a) the neeps would not all be down; (b) if they were down they would just be coming to the hoe; (c) the bull might break out; and (d) Roma was due to have a calf, and if we weren't in attendance she might have anything from twins to after-calving paralysis.

Mind you, I quite see the grieve's point of view. So many things can happen in a day, as all farmers know. And not for the first time I yearned to turn the whole thing off at the main.

However, like all other farmers in like predicament, we're going to the Show just the same, and we'll love every minute of it.

Surely for a couple of days the neeps can't make such awful whoopee, and Frankie the bull must just stay shut up in his house, while our expectant mother must take a chance. Who does she think she is anyway?

MARRIAGE OF COUNTRY CONVENIENCE

24 June 1961

The counties of Moray and Nairn are, for the purposes of administration, united under the name of Joint County. This is a useful arrangement, but, of course, everybody knows that it is purely a marriage of convenience and that we are unequally yoked together. Thus we pursue our own private lives amiably but inexorably, accepting the fact that oil and water cannot mix. Moray is all civic virtue and common sense, with as keen an eye to the main chance as any shire in Scotland; while Nairn, a good deal less wealthy, is inclined to think it virtuous to have a mind above filthy lucre and to find the more romantic Highland tradition more congenial.

Wayfarer
The highways department set up signs between Moray and Nairn to tell the wayfarer when he leaves one county and enters the other. This seems superfluous, because anyone with eyes in his head cannot but see the different set-up a few miles after he crosses the Findhorn outside Forres.

In Nairn every other house dotted along its picturesque roads (which, incidentally, have a better surface on the whole than those in Moray) carries a nice little white sign saying "Bed and Breakfast." Moray, rather looking down on the tourist industry, puts up Dutch barns and irrigation systems. If it must come to terms with the visitor, it will do it more discreetly; which is a pity, because the food in Moray is excellent.

The people in Nairn speak with a different accent from their neighbours in the adjacent county. Somewhere in the nine-odd miles which separate Forres from Nairn men decide to say "Nern" (with a soft Highland "r"), while back in Forres they are still saying "Nayrrn" with a good rolling Lowland reverberation. The linguistics department of Edinburgh University are doing some research this summer in Elgin, so I hope they find time to tackle this intriguing phonetic puzzle.

That we accept one another's differences so genially might seem a measure of the mellowing power of time and propinquity, but in the event we turn out to be not entirely mellow. Ancient memories of which we were barely conscious assert themselves, and we find ourselves suddenly aware of being at one with ancestors who thought of the Findhorn as the natural watershed between two counties, and a boundary that must be respected.

Status Quo

We had a meeting the other day about sending eight Moray pupils to school in Nairn. There were only two miles difference between the different transport arrangements, but for the pupils' parents it might have been 200. The appalled incredulity with which they greeted the proposal was so impressive that for once authority hastily retracted and resumed the status quo. The Moravians thus contemplated a victory, due not so much to their indignation as to the compulsive power of the history and geography engendered by the loveliest of Scottish rivers, the Findhorn.

The Findhorn has always exercised a curious effect on the inhabitants of her valley. In pre-historic times a large population lived here, sustained by hunting and farming. Shadowy legends remain to tell of their beliefs and rites. There are innumerable cairns and forts to remind us of the perils, which must everywhere and at all times attend the human state.

Later, Columban missionaries came with their skills and elevating faith. On the minds of the indigenous Celts these men must have exercised a wonderful excitement, at once intellectual and imaginative. To this era in Findhorn's history we owe the collection of exquisite carved stones and crosses which ornament its valley. The tradition of "pietas" is not extinct.

AFTER THE BALL

29 June 1948

We're feeling fair off our stot since the "Highland" is now over. However, there is a hillock of work waiting us so we have to get on with it. Only we'd not be human if we didn't spare more than a glance at backward looking.

The grieve, who is busy topping a grass field beside the bridge, thinks of all the fancy sweeps, swathe turners, and buckrakes that stood splendid in blue and silver among the machinery in Inverness.

Who can blame him if he regards our home-made cowp cart with a curl of disdain when he thinks of the Tip-it-als and three-way tippers that flaunted it in the machinery-in-motion section in the great tree embowered park beside the river?

Sour-grapishly, we pretend that the precision machines would never do amid our rocky gradients, but if a fairy godmother were to say …well…

Firmly, I take a hold of myself. I better confront my desk. It's a sobering sight. Come on Elizabeth, get on with it.

Here's a chit to say we have a quarter cwt. of protein left over from the last coupon quota. It's hardly worth bothering with it, but by the time winter is here it will be more useful than gold. Better take it up.

What's this now? The manure bill from one firm, and, oh heck, the other firm has mixed up the coal account in their one. Phone and say please send manures separate from coal so that I may apply for the marginal land grant in a straight-forward fashion, also send note re lime application.

Look at this flossy catalogue now. Every implement well over a hundred pounds. Instead of bunging it into the waste paper basket, I guiltily file it, not quite admitting that I'm keeping it to yearn over at a more appropriate date.

Now come three accounts hard running. Where the mischief is my cheque book? A hasty calculation over my illegible stubs shows, surprisingly, I have some money in the bank.

I pay up and reflect that the Wool Fair is next month, so we should have another tuppence or thruppence to get by on then.

Three personal letters which'll get answered sometime are succeeded by a fantoush card from the Ministry of Fuel and Power to say there ain't no petroleum board any longer, will I please make suitable arrangements, therefore.

My pen falters and I am seeing Inverness looking like a northern Camelot. This won't do. See me over the typewriter.

MIDSUMMER MIDNIGHT

My bedroom window opened wide on the summer night, and lying in bed at midnight I could contemplate the blessed shadows, pansy bloomed, that softened all the contours of a summer day.

An earlier shower had compelled the birches to release the burden of their sweet insistence on the quiet air where a solitary white owl blundered drowsily among the drowsy moths. The lovely night was lapped in Lydian airs when suddenly a horrid stridency annihilated calm.

I leapt furiously from bed to murder the assailant of my peace. "Where's your decency?" I asked, sticking an indignant head out.

"Hullo," said a hoarse, familiar voice, and there was Duck, who, not content with the ordinary din in which she lives, was adding to her own duck-bred noise by drumming with an obstreperous bill on an empty square-cut tin.

"I thought that would take a trick," she continued with satisfaction. "It's too quiet hereabouts, too damn quiet in fact. Look at that." She indicated with a contemptuous neck the still form of Drake, who was sound asleep at the henhouse door. "I've asked him till I'm tired to come out and see a bit of life, but oh no, 10 o'clock and it's bye-byes for Drake."

No-one could fail to be struck by Duck's disgust.

"Come 10 o'clock I can do with a bit of fun and there's minnows in the burn." Her eye swam in glossy greed. "And froggies, too, that play leap-up-and-over among the forget-me-nots. The minnows hide behind the marsh marigolds, but I know where to find them." Saliva drooled anticipatory from her plebeian bill.

"What about the eggs you were supposed to be sitting on?" said I.

She slid her bill down a prettified right wing feather and gazed into space.

"Yes," I continued, "and if Mary hadn't found where you rolled your eggs in ones and twos beneath a bit of board, we'd have fallen for your tale that rats had taken them. Fie for shame, horrid bird."

A smirk illumined Duck's leering visage. "You know what men are," she murmured, casting a languishing glance at her sleeping spouse. "No resources in themselves, and with all these unattached hens floating about."

She shrugged with liquid grace, and then, to close the subject, "You're not coming down to the burn then?"

Hastily I slammed the window on the siren night and its shameless priestess. As she waddled off I heard her intimate, fleering laugh and it pursued me down all the dim avenues of sleep.

EXILES, HERE AND THERE

We are daily now expecting the notice which will send Frank back to Italy. Several of his friends have already come to say good-bye, their eyes shining with such joy that one cannot restrain a pang for all the exiles who toss about Europe in these distracted days.

It is true that many of the returning Italians are not without doubt and anxiety about what they may find when at last they reach journey's end, but no anxiety, however wry, can mar the deep, abiding joy of getting home.

So not for many more weeks shall I in imagination walk round the village square of a northern Italian hamlet I have never seen nor am likely to see. Regularly after byre-time Frank is my guide past the church where the Madonna dispenses sweetness and holiness, past the shops which crowd the street outside, past the houses.

Here is the greengrocer, here the perfumier. Round the corner is the house "where my lawyer lives who does all my personal business," and here "my casa stands." We watch the women washing clothes in the river and hear the children laughing as they learn to swim. At night we stand alone in the dusty street whitening under the argent moon.

As season succeeds season we hear of what is now happening in Italy. Grapes and peaches ripen in the hot sun, and slow oxen work in the fields. So often I have heard of what one must do in planting sweet potatoes and hemp that on a frosty spring morning I have felt a vicarious anxiety for crops which for all I know may not, now or ever, be planted under the blue Italian sky.

Memory sharpened and made bright by the pain of absence calls up for all exiles the dear intimate things which unknown to us create the beloved place for us. What small things make home for us voyaging mortals! A shelf of bright dishes above the kitchen fireplace, a handful of wild flowers, the shape of a hill rising from a sunlit sea. That for us is home. The small part of the "too much loved earth" that each of us possesses is made the more lovely by separation.

As Frank prepares to go home our neighbours who have a son newly gone to Italy seek to find out about the foreign land. In halting English he tells them of a country which in essence after all is not so unlike ours. It seems that all the world over men must wander and women look for their return, mothers yearn for their circling children and Ulysses for ever turn his face towards home.

I wonder if ever our young Scots lad will meet the returned prisoner. If they do, how strange to hear from foreign lips that the river still sings its song to the church, the beech leaves still lie in rustling copper swathes in the still December air, the old home still listens for the return of its dear exile.

WAITING FOR THE WORLD TO C OME ALIVE AGAIN

1 July 1961

Intellectuals wait for Godot but farmers wait for the weather. At least that is what we and our like have been doing all through nearly thirty droughted days of an unkind June.

Up here, where the soil is thin and poor, we depend on a "misty May and a drappy June" to get our crops moving; but the last three months have been so deficient in sap that even in the fertile Laigh, where the depth and recuperative power of the soil can counteract the drought, farmers have had to turn on the artificial irrigation to keep heart in their pasture and corn, while hasky wind blasted the early potatoes and miserable wee turnips could hardly lift a pitiful finger out of the drill.

But the Laigh can always survive. At the end of the sixteenth century, when there was famine all over Scotland, the now buried estate of Culbin in the Laigh exported corn to the rest of the country. Meal rose to 30s a boll (a boll being 150lb) and after that it had to be transported over the Grampians. We never had it so good in the high parts so we are inclined to gloom when a parching east wind racks us daily during our short season. There is nothing very heartening in watching whirlwinds begotten out of nothing dancing their dervish way over the brown fields. True they are but children of dust but it is expensive dust when much of it is the expensive fertiliser we sowed on the surface a week or so ago.

Soon a neighbour as sorely driven as ourselves is sure to come along and remind us of 1829, when after three months of solid drought and heat the rain began on August 2 and fell in sheets not in drops, and "there was a peculiar and indescribable lurid or rather

92

bronze hue that pervaded the whole face of nature as if poison had been abroad in the air."

You hang about waiting for rain. "Nothing happens, nobody comes, nobody goes. It's awful." So it is but there is a remedy.

Silage Pit

When all hope departs of ever filling the silage pit and you are calculating how many cows you can afford to keep on nothing during an ageless winter, you must rise above it and get out the mower. Bolt it with as cheerful a stramash as possible to the tractor and set off to cut whatever miserable grass you have. Briskly ring your neighbour who does your hay bailing on contract and tell him that you are making hay. Nothing will happen till the hay is all cut and then next morning after a drying night the first tentative drops of rain will fall.

Whistling nonchalantly, make your way to the implement shed and haul out the hay rake and get going. After an hour or two when roughly half the hay has been turned and swathed the rain will fall in earnest, but you must still press on so that by the time the baler arrives in the field everyone will be running about in oilskins and the hay will be soaking up the downpour like a sponge.

Drastic Action

It is as well to recollect that this drastic action on your part may not be universally popular throughout the district. Some folk who have decided not to wait for the weather may have made up their minds instead to go for a picnic or on the rural outing. Straining to see through rain-streaked bus windows the picturesque and romantic scenery of upper Speyside they may be inclined to take a correspondingly dim view of your behaviour, and may not share your delight in seeing the grass getting hold at last of the high nitrogen fertiliser which will boost it into high green billows which the wind will lick to silk like a cat cleaning her fur.

Holiday makers plowtering in plastic macs and plastic sandals cannot be expected to participate in your farming glee either. But the briar will shed her petals and her fragrance and your world will come alive again.

CHANGING A RURAL CONSTITUION

8 July 1961

There are benefits, invisible but potent, in being an oldest inhabitant in a Rural committee meeting, as I found last Thursday when I attended a village hall committee which lasted all a Scottish summer eve from half-past seven till a quarter to eleven. I am not really the true ancestral voice of the glen because there are folk who farm here the places their forefathers worked hundreds of years ago; but such people had more to do on a drying night in the middle of the haymaking season than come to the hall meeting.

Local Endeavour

The reason for the meeting was a proposal to change the constitution. Our hall was built by local endeavour in the 1930's and it needs now both running repairs and extensions which we cannot do on our own. We can get help from various authorities but we must have a constitution of which they approve. This we do not have, and to bring ourselves up to date we must make some new rules.

Our present rules were more evolved than drawn up by the chaps who actually built the hall and they have acquired a kind of ancient significance. When anyone suggests changing them, the committee shriek "Sacrilege!" and, like Boadicea, "seek with an indignant mien, counsel from their country gods." But, after all, 30 years is a long time and even in darkest Dava time marches on, so that most of the diehards are now gathered to their Valhallas; yet a core lives on, traditionalist to the end.

And on Thursday the traditionalists were there, all the four, against eleven newcomers, of whom I was one. However, having been brought up to do my homework, I did my history – 21 years of it, the years I have lived here. The traditionalists had an inherited mystique but I had facts. This was wonderfully effective and I began to think we'd get home before midnight.

But the advantages of age and local knowledge are not of themselves sufficient to ensure swift victory. One should never overlook the fact that at meetings in rural places people do not say what they mean. When they argue against the dramatic club electing a representative to the new committee because the dramatis personae are the same as in some other club, they are not in fact saying that it is thus possible for one person to have two votes. They are expressing fury at the choice of play the president made.

Dissident Parties

When the traditionalists of the hall committee are also members of dissident parties in all the other clubs who have a right to elect representatives the proceedings assume a Medusa-like fascination for the onlooker – half double think, half Ivy Compton Burnett.

Old inhabitants who know the feuds raging in rose-smothered villages must resist the temptation to admire the child-sweet voice of a secretary questioning legal rights when she is really saying, "Over my dead body will you elect another office-bearer I dislike."

No one of course believes the youth alleging that his interest in the whole affair is on behalf of his forefathers is doing anything other than kidding himself on that he can sway the multitude – all the 15 of us.

In this atmosphere silences can speak louder than trumpets. Anyone looking like one of the maturer goddesses and wearing Harris tweed with the untroubled air that goes with complete confidence in one's own efficiency can by the inclination of her silver head make or mar any motion to do with the Rural.

But, after all, the crux of the business is whether we shall get money to buy new chairs and re-do the Ladies. If we want the siller we must adopt the new rules. Ancestral voices prophesying woe fade down time and we sign on the dotted line while the haggard chairman reels with fatigue.

..

WET, COLD AND MISERABLE

20 July 1948

That's me, and no wonder, for I'm just newly in from feeding the chickens, and the latest lot who are still with the hen chose to spend the wettest night of all the wet July in the open, although the door of their hutch was wide open and the hutch itself beside them. What demented brutes hens are!

I spent an exasperated quarter of an hour chasing them into the dry, and all the time my Wellingtons were slipping helplessly on the wet, slooshy wood that makes the step into their house.

All the time, too, the rain gushed down. Every gutter and barrel about the place is overflowing with the flood. On the hill the burns rumble and toss full and gorged with the rain that shows no sign of easing.

The moor is slashed with the white foam of their tempestuous journeying, and the heifers, who spend the summer on the hill, are coorying for shelter in the kindlier hollows whose backs are to the driving storm.

At home the tattie and neep drills are up to their necks in water. They say the tattie inspectors started yesterday looking at the earlies down the country. Poor wretches, I don't envy their ordeal by water.

Imagine spending to-day wallaching up and down guttery potato drills! It can be only scant comfort to reflect that the foliage of earlies is only knee deep and not waist deep like half of the main crops.

Instead of writing this I ought to be in the scullery boiling milk for Daphne, our youngest calf, who has a pain in her pinny, but my ear tells me there are plenty folk there already, all wet, all needing things, and all barging into one another.

Listen. That's Mary shooing the dogs out and telling them the barn is the place for beasts with wet coats and muddy paws.

All that screeching is my eldest daughter menacing the hens who are determined to spend the morning in the porch attacking the pail of dinnertime mash.

The triumphant squawks come from infant number two who has discovered an egg in the car.

Here's the grieve demanding to know the whereabouts of the disinfectant I stole out of the cupboard in the shoppie and forgot to return.

Losh me what a sotter! I wish I was a duck. Look at ours turning satiny somersaults in the pond with well-oiled ease.

It's a good thing something likes this weather.

PAINS AND DISEASES

23 July 1947

No, none of us have got them, unless you count small black Belinda, Bella's surviving twin calf who overate herself yesterday and has been on a diet since, to her vociferous disgust. The pains and diseases I am thinking of are the ones our potatoes may or may not have.

Until the weather broke (which it did today with a vengeance being all north wind and haar) I have been spending my leisure moments among the Edwards and the Di Vernons trying to keep them up to stock seed standards.

Growing seed potatoes is a fiddley business, for you must go through your crop twice and remove all diseased and not true-to-type plants.

Although I do this every year I am not in the least confident of my ability, and spend my mealtimes seeking the flat of that invaluable booklet called "Maintenance of Pure and Vigorous Stocks of Potatoes" which the Department of Agriculture publishes. Sometimes, however, I am inclined to doubt the assistance of this manual, because when I read of the viruses which can attack the potato I begin to think my seven acres are positively oozing disease.

"Susceptible to viruses A and X and has some resistance to them, also to virus Y. Mosaic diseases are not obvious and are difficult to diagnose. Susceptible to leaf roll and to viruses B and C." Think of all that labouring in the heart of that sweet girl Di Vernon.

Of course, after I've laboured up and down a drill I forget about the viruses, and having got my eye in I can pick out the transgressing plants easily.

A good, whacking mosaic looks up at you evilly and grins, while a leaf roll turns up its little withered hands in the most despairful gesture imaginable. So I howk them out, tubers and all, and shoot them out over the ravine which borders the field.

If the light is good – that is dullish and undisturbed by blinks of sun – and the weather calm without a wind to ruffle the foliage, rogueing is a pleasant job, though one that calls for a high degree of concentration.

But although I have no eyes for anything but the crop, my other senses are companionably aware of the gently burgeoning countryside, of the scent of briar rose, and clover, of the warning cries of curlew and plover. Far away I can hear a train to remind me that the islands of peace are accessible to others than myself.

At last it is too late to see, and I discover that my dungaree knees are wet and my Wellingtons harbour nyattery granules of earth in their soles.

FOR SALE, A FARM

3 August 1948

We went for a motor run one golden summer afternoon up a forgotten Highland strath which leads precisely nowhere, for its ever narrowing gorge finally debouches into empty moors overhung with shadowy crags.

The road which connects this place with the outside world has room for but one car, so that when you meet another you have lots of manoeuvring to do. On one hand there is a high deer fence and on the other runs the river gold and glancing in the sun.

Now and again the land widens slightly and men have made crofts there, but the glen is not suited for agriculture and has very sensibly been taken over by the Forestry Commission, who are already beginning work there. But you would be wrong in supposing that there is no agriculture here at all.

Just before the road becomes a defile the country broadens out appreciably and at its widest span there is a farm, no, I mean a FARM. The proprietor of the FARM has died and now it is for sale.

My friends, who are tolerant of my eccentric occupation, urged me to go and have a look round and, of course, I was delighted. We went down a drive where blossoming shrubs elbowed their fragrance into the warm and sleepy afternoon. We found an old grey mansion standing before obsequious lawns all smoothly green. There was a garden sunk in gold saxifrage and mauve violas and then we came to the Farm.

Aristocratic cows dallied with white clover in the rich fields and shrugged their svelte shoulders at the sight of plebeian strangers. I thought them a bit uppish, but when we came within sight of the steading I understood. I just wish I could describe that steading.

Everything was tiled in glossy blues and reds, greens and whites. The cow stalls were tubular steel and resplendently silver. There was a washhand basin with hot and cold water for the dairymaids to wash their hands. I shouldn't have been the least surprised to have seen set-in tubs for shampooing the cows tails. Don't tell me about bathing pools in Hollywood. They had nothing on this set-up.

Besides all this magnificence there were implements and other things all on the same scale. Also there was a tower silo. None of your poor relation pit clamps here.

One of my friends suggested I should put in an offer for the place. But I knew I should not even if I had the money. Fancy having your yokelness rubbed into you by yon flossy steading.

CONNOISSEURS OF HONEY

5 August 1961

An acquaintance has just been in to ask if we'd take in his bees for the heather honey gathering. Down in the Laich where he lives the clover season is over and he must find fresh meads for his indefatigable charges. Most low ground apiarists in Moray send their hives to the moor in early autumn, and, since the journey entails much anxiety and arrangement it is well to have the spot where the bees are to be situated looked out well in advance.

Beehives cannot be dumped anywhere on a moor, be it never so flowery. For one thing the chosen spot must be reasonably accessible; one does not willingly career over a moor with a lorry load of hives however securely one lashes them down with ropes. It should also be borne in mind that such expeditions take place at night when the poor hardworking bees are all home from their labours, so, when you add darkness to the other perils of the journey, accessibility is all the more desirable.

But bees themselves have ideas on their new home. Acres of flowering heather are all very well, but not if you have to go out in a gale to gather honey; and if it is an east wind gale then you think it better to stay at home and warm your feet in the honeycomb. It is an unfortunate fact of nature that here in upland Moray wind seems inseparable from the moor so that the beekeeper must go poking around in the lythe of plantations looking for the ideal coign of vantage.

Even then his worries are not over; he must cast a wary glance round at the sheep who imagine they have a prior right to the hill. Every countryman knows the controversy that rages about whether grouse and sheep agree together or not; but the enmity which exists indubitably and sadly between sheep and bees is less well advertised.

A beehive exercises an appalling fascination on the sheep. She smells it, pokes it, and finally, after reducing its inmates to hysteria, she turns it over. Instead of suffering the just retribution she so richly deserves her woolly coat protects her, and she is able to complete her work of destruction by clawing her malicious back and sides on the nice sharp edges of the ruined hive.

Innisfree

The more one thinks of beekeeping the more one is forced to the conclusion that Yeats was all wrong when he thought how soothing it would be on Innisfree to keep hives and beans. I sometimes wonder, by the way, what kind of flavour his honey had.

Here in Moray, since we have a tradition of beekeeping that goes back to medieval times, we are connoisseurs in the flavour of honey. Clover we rate as rather ordinary. It tastes too much of plain sugar and it oozes too readily on the plate. Heather is different. We like its aromatic fragrance and its firm texture. One smells honey rather than eats it, which is why we have a liking for honey produced from lime trees. In addition to its almost fruity taste it has a lovely pale green colour to commend itself to a finicky appetite.

I remember long ago in the South West of Scotland being given honey distilled from the sea pinks which grow in such profusion there that one might think the dawn had fallen to the sale sea marge. This honey was at once sweet and briny with a curious after gout of iodine which made it titillating to a palate more sophisticated than mine then was.

But wherever the honey comes from – Hymettus, Dava, or the Solway – let me eat it from the comb from which it "droppeth sweeter far" than from any press or extractor or whatever modern mechanical gimmick gets rid of the wax.

"LOUP ON, YOUR MAJESTRY"

12 August 1961

We are all agog with delighted excitement about the visit the Queen and Prince Philip are paying to Banff, Moray, and Nairn on Monday. They are arriving at Macduff on the royal yacht, and we are determined that we are going to hang out every flag and furbelow we can, not only to welcome them but to erase the memory of the way we behaved the last time we had a monarch visit us from the sea.

That was when Charles II came over from Holland and landed at Kingston, at the mouth of the Spey. The vessel which brought him could not come into the harbour, so a boat was sent to land the King. "This boat could not approach the shore sufficiently near to admit of Charles landing dry shod; whereupon a man of the name of Milne, wading out into the tide, turned his broad back to the King and quietly bade His Majesty 'loup on'."

Charles needed a little persuasion but did eventually come ashore. Then he was hurried to a house in Garmouth and there compelled to sign the Solemn League and Covenant while a puckle Scots gentry stood round bleak and uncompromising. Thus, with little ceremony and less hospitality, Charles set foot in the country of his ancestors. No wonder we want to forget all about our behaviour then, and are happy to think how differently we order things to-day.

Naturally we are hoping that the weather, of which we are apt to be a little vain, will live up to its Moray Firth reputation and that the sun will recognise his loyal duty to shine over the calm sea and gild the whole of the royal route which runs from the jostling cheerful little ports through the fertile Laich on to Nairn, where the moor road will bring our visitors to Grantown, superb among the mountains. Each of us thinks privately of some particular view we'd like the Queen to see but on the whole we are willing to concede that the authorities have done a reasonably imaginative job. Perhaps they could hardly go wrong in such a country.

The earldom of Moray has had a long connection with royalty and few earldoms in Scotland can boast of a bede roll of names more eminent in the annals of their country. Randolph, first Earl of Moray, Black Agnes of Dunbar "The Good Earl of Moray," "The Bonnie Earl of Moray," are household names in Scotland. Perhaps the most powerful of them all was Lord James Stewart Earl of Moray and half-brother of Mary Queen of Scots.

Queen Mary

It was in his company that Mary came north in early September in the year 1562. Like Queen Elizabeth he came to Elgin from the east, and her route also took her from there to Kinloss. Queen Elizabeth is to pay a visit to the aerodrome there, and we hope she will like the place as well as Mary did.

We read that she was delighted with the accommodation provided for her at Kinloss. She stayed at the abbey for two days and enjoyed the gracious and ample hospitality of the cultured churchmen of the day.

An old menu set out a bill of fare used for the royal visitor, and for dessert it includes "apples from the chanter's garden." Robert Reid, Abbot of Kinloss and a commissioner at Mary's wedding to the Dauphin, was one of the most eminent men of his time. Among his many accomplishments he numbered gardening, and it was to him Moray owed her wide orchards with their 146 different varieties of apples.

Our royalties are to be entertained to lunch in the beautiful county hall in Elgin. This is a room full of dignity and light and suitable in every way for the occasion. We hope that those who planned the meal will follow the example of the old monks in Kinloss who had the sense to include local delicacies in the banquet.

Above all, of course we hope for a truly royal progress and a happy holiday at its end.

WHEN THE CAT'S AWAY

17 August 1948

This is a slack time of year on the farm for the hay and hoe are finished with and as yet the hairst is a fortnight away, in these parts anyway.

There are small things like drill-harrowing neeps and rogueing tatties for the final stock inspection, but on the whole we are at ease. So we think it quite a scheme to attend the many local agricultural shows which take place just now.

You can see us ambling ruminatively about at Grantown or Elgin or Keith, and looking at us about twelve noon you'd say you never saw a more serene or placid people.

Slow, amiable, and country looking, we greet our friends, and chewing an end of grass we breathe contented "Aye, aye's" at one another.

But as the afternoon wears on the spectator of the country scene begins to detect a certain distrait air among us. Our eyes grow worried, our brows furrow.

Finally about five o'clock the tension reaches fever heat and we hurl ourselves into our chariots and make for home with all despatch and considerable shriekings of tortured engines. As a people we farmers have the most revolting attitude to machines.

If you were to ask any of us why we rush off like this we'd tell you in one word, "the bastes."

Between the farmer and his bastes there exists a strange, deep bond. As long as he remains at home the bastes remain quietly behind their fences eating their grass and only raising their grateful heads to say "Thank you, dear master, for this lovely grass. The clover is luscious, and since you topped this field the flavour is superb." Idyllic.

But the minute the brutes read the papers and see the advertisements for the local shows all is changed.

Then they keep their ears cocked for the sound of the car going down the road and then, "All together, girls, let's make short work of this fence."

The farmer knows this in the abysses of his subconscious, but for a long time the knowledge does not reach his senses. Quietly he goes on with his modest pleasures.

At home the crucial fence-post goes and the wires part with a high, metallic scream. The farmer feels restive. The dear cows walk forth into the neep field and comment insolently on their winter keep.

Graciously they devour wallops of tattie shaws as they meander down to the burn. Deep chords of alarm now are sounding within the farmer's breast.

By the time Cathy the Cow has summoned her calfies to play hide-and-seek in the ley corn, the farmer has become so uneasy he can stand it no longer.

"Home already?" asks the cows impudently as they watch him struggling into his dungarees and whistling for the dog.

MACGREGOR GROUSE AND THE CAPER

19 August 1961

Everybody interested knows by this time that there are no grouse on the moors this autumn, and therefore guns are out of place in a season when the cold weather of early summer killed off the promising coveys of a warm and premature spring. A few old birds, of course, have survived, and these, because of their scarcity and their vigorous and savage air, are on the way to becoming personalities in their own right. We know that neither our laird nor his party would have the effrontery to lift a gun against them, and Macgregor the cock grouse who rules our eight hundred acres of moor, is safe.

Macgregor leads a life of solitary power, unmolested even by his hereditary enemies, the foxes who live in the sandy hills beyond the kettleholes. When we go to turn the sheep from the hill he raises a scarlet and choleric eyebrow but says no more than "tut tut." We like to think he is indulgent to our trespassing because he is grateful for having had the use last winter of our nearer grass park, where we were in the habit of scattering corn to our few hens. Macgregor put his pride in his pocket then and stamped around in his big feathery feet eating along with the plebeian farmyard fowls.

He does not, however, allow Bob, the collie, to have any liberties on this account, and when Bob inquisitively noses him awake from his after-dinner snooze Macgregor rises straight up into the air shouting like an abusive alarm clock.

We are often surprised at how domesticated wild game birds can become. Some years ago a capercailzie took up his winter quarters in a hen house down the road from here. Capers became extinct in Great Britain in the middle of the eighteenth century but were reintroduced from Sweden in 1837, since when they have flourished and spread widely over the pine wooded regions of the Highlands.

The caper is a very grand gentleman indeed – large and handsome – but a dreadful show-off in front of females. He bawls out a loud double cry summoning all the women to come and see him. While he thus vaingloriously shouts he puffs out his plumage and unfurls his tail and generally lashes himself into such an ecstasy of vanity that his eyes literally disappear in his head.

Our caper, who was known as Maclusky, was just a braggart, as the rest of his clan. Indeed, love combined with insensate vanity were nearly the end of him because he stood under a tree shouting out how wonderful and beautiful he was till his eyes were blinded by his swollen head and a schoolboy was able to pick up the besotted bird, and would have carried him to his ignominious doom if chance adult counsels had not miraculously appeared and made themselves prevail. Maclusky was therefore released and for all I know may still be leading his conceited life out in the deep forests of Darnaway.

Sport on the moor does not depend all that much on numbers of birds. An observant eye, a pair of strong ankles, and a digestion that is satisfied with a piece of oatcake and a lump of cheese for a meal are all that one needs for pure enjoyment on the hill.

I often think when I see elaborate preparations for a shoot, how lucky Charles St John, the sporting naturalist, was when he lived among us in Morayshire. In those days in the early part of last century "The Season" had not yet assumed its social and economic importance on the Highland scene, so that St John was able to pursue his enviable life as a sportsman and naturalist anywhere he liked since any Highland laird would then give royal hospitality to a reputable gun.

The Allan-Hays who wished to be known as the Sobieski Stewarts were others who enjoyed such country pleasures. Alas, that their expressions in print of these happy pursuits should have led to their sad exposure!

IN TIME OF FLOOD

24 August 1948

After the rain of last week I now have every sympathy with Noah's dove. What a disastrous time it has been with day after day of heavy, unrelenting rain.

Ditches, ponds, rivers and dams were early gorged with flood waters and eventually were themselves submerged under ugly, brown torrents which seethed and tore their destructive way over the unhappy countryside.

Today the worst seems past and we were able to go to the byre in the morning without getting soaked, although the cows had to be rubbed down as usual before milking to prevent the trickles from finding their way down the beast's flanks into the pail or on to the milker's forearm. It is still far from dry, however for thin, hissing showers of vicious rain needle away on the corrugated iron roofs.

All the same, the dry intervals do prolong themselves and we begin to hope that at last "the fountains of the deep and the windows of heaven" are to be stopped.

The river shouts away hoarsely but the ditches and ponds are slowly subsiding. But, oh dear, what a mess there is everywhere.

We all wear Wellingtons and go our oozy way about the muddy tracks we call farm roads. Poor Mary fights a losing battle with muck and in spite of commonsense continues to scour the scullery floor.

Clotheshorses and pulleys are full of dripping garments, while drowned oilskins hang on the pegs and nails in the porch.

The grieve's extensive wet weather wardrobe gave out yesterday and he appeared in a natty corn bag worn cornerwise over his head and draped down his shoulders like a somewhat grimy member of the Ku Klux Klan.

The beasts were as meeserable as the humans and sought shelter behind an old straw sow. They kept wonderfully comfortable there, too, but the sow is looking a bit tired.

But for all our moans and groans we realise we have little here to complain about. I was at Forres yesterday and when I saw the bonny fields just coming to hairst I could have wept. You'd have imagined the sea had swept over them.

The sight of the advertisements adjuring the public to lend a hand on the land make us smile wryly. Happy hostels, ample fare, good pay and travelling expenses are all offered as an inducement to help with the harvest.

Nothing is said about the torn and battered fields that must be garnered. And if they aren't, then folk, we all go hungry. Not nice.

COME TO THE FAIR, THE FUN AND GAMES

28 August 1961

Two things for us emphasise the official terminus of summer; the appearance of new school bags in the shop windows and the day of the Nairn Games. Let other pens dwell on such odious matters as the end of the school holidays and their corollary of home lessons for the children, and the hardly less onerous task for the parents of buying replacements for outworn school uniform. Wet and miserable though this August morn may be, I prefer to dwell on the games.

The Nairn Games have always enjoyed unprecedented popularity in our part of the world, though they lack the social appeal of the really big games events of royal Deeside. But Nairn is easily reached by a population living on the eastern seaboard of the Highlands and besides it is a pretty crisp little town and pleasant to visit.

Enchanted Hills

Of course it is wonderfully lucky in having acres of links which form a naturally green amphitheatre whose potentialities generations of tourist-minded citizens have set about developing. There is nowhere like the Nairn links with their background of firth and enchanted hills for dancing a reel or running a mile or just for meeting one's friends.

But though the scene is so magically set, the games of themselves would lack verve and excitement if they were not skilfully pointed up by the fair-ground which surrounds them. For at least a week before the day our already crowded roads are further crammed from verge to ditch with the traffic belonging to the show people. Shuttered gorgeously and bedizened like oriental beauties, the living-vans surge along behind the flamboyant modern cars which move them from place to place. Behind them come the heavier vehicles grandiloquently advertising the chairoplanes and dodgems racked tightly within their blazoning walls.

There is something so arrogant and overwhelming about the long cavalcade that all traffic gives way, and cars with boats lashed to their roofs, and waving red flags fore and aft stand meekly aside, thinking they will reach Findhorn more conveniently when all the shows have passed. As for the travelling fair, once it begins to smell the first iodine breeze from the shore it loses a little of its ordered progress and like Xenophon's Greeks hurries to the waves shouting between its snorts of diesel, "The Sea! The Sea!"

It is the custom of the nearer country folk to go for at least one night to the showies and do the shooting galleries and all the thrills and spills. For once they have time on their hands with the hoe over and the hairst not yet ready for the combine. Countrymen with sun-bleached eyebrows and skins as sunburned as the gipsies who lure them from one marvel to the next, end the evening with foolish hats on their heads and their arms full of silly dolls.

Above the soft crash of the waves tinned music shrieks louder than the garish lights which outline every stall; and above the music rise the flaunting voices of the neat swarthy-skinned folk who call their wares and excite custom, their barbaric ear-rings and necklaces bright as their own predatory eyes.

Then at last it is Saturday and the day of the games. If you have not done your week-end shopping on Friday then you are lost – for not as much as a stale soft biscuit will you be able to buy in Nairn. Every shop is closed and every road leads to the links. Here athletes meet and serious spectators gabble about record times. Little girls huddle in coats to protect their finery against the Scottish summer. Pipes skirl and you meet old friends and make new ones. It is all picture post-cardish and gay. To-morrow is autumn.

BOOK GAMBLE

My annual library subscription is due and, as usual, I ponder on the great ingenuity necessary to get good books in the country.

Of course if we did not live in Arcady we could have an electric wireless or even television but a battery wireless, irrevocably thirled to the Home Service, is not very satisfying entertainment for the whole of a long winter. Books are necessities.

A subscription library comes in very nicely though perhaps more so in theory than practice. If I could get down to see its shelves it would be pleasant but, as it is inconvenient for both of us to be away from the farm at once, I send my returned empties by post.

I also include a list of books I want to read. Sometimes I am lucky. More often some one has just borrowed the book under the very nose of my returned parcel.

The librarian then gallantly looks for something suitable. She is a dear good child but I wish she did not think that because I live in the country I want to read about the place.

Reading about hills and moors is quite different from living among them. As for sunsets and dubs, I've just to open the back door and there they are, triumphantly unbookish.

Rural Discomfort

I dislike extremely books about how brave it is to be funny about discomforts. There are too many undisciplined elements making for damps and draughts in rural life and the proper approach is not humorous but to draw practical plans for banishing them.

I enjoy reading about successful farmers who have made such lots and lots of money that they can build roofs over the whole farm – nearly – and do their work there comfortably and to pot with scenery and weather.

I like reading about the mannie who made a poultry ark for £4 and a paragraph on scour in calves is enthralling.

Erleigh on the history of agriculture is also supremely satisfying.

The Carnegie Library is very helpful to folk like me. If you are not impatient you'll get what you want. The trouble is that sometimes you don't know what you are asking for.

The local post office and I won't soon forget the winter I got entangled with Lecky on the eighteenth century. Sixteen volumes.

Good books in the country would then be thus. Either you know what you want, and don't get it, or you don't know what you want and do get it.

COUNTRY CONVERSATION

How delightful it is to listen in to country conversation, especially that carried on in the more isolated parts of Scotland. Every county, every district has its own flavour and yet all the talk remains essentially and joyously Scots, however it may vary in detail.

The other day I ran into a lady whom we call the Duchess. The Duchess, who hails from Buchan, had newly returned from visiting her newest grandchild. She was glad to be back, she said. "Because I'm auld and when a body's auld they like quiet at e'en."

"Quiet at e'en," the words ring in the heart and at once old age becomes what it should be, something infinitely desirable and serene. Conversation with the Duchess is like the country she comes from. It is usually shrewd, always entertaining, and occasionally it brushes shoulders with poetry.

One of her many grandsons was once passing the time of day with us. "There was aince a darkie dwalled ayont the hills of Ennory," he began. The rest of the tale was lost in commonplace, but after years I still remember the irregular but unmistakable trochees of his opening, and I still speculate on the odd adventures the darkie must have had.

The speech of the peasant in these lower hills of Scotland has a quiet and minor cadence like the character of the people who use it. It is patient rather than dramatic, and it is above all accurate. It has its feet very firmly in the earth, but one can still hold a plough and watch the peewits whirl in the enchanted spring.

To go from conversation here to talk in the Highlands is as bewildering as a somersault. Form becomes now the main consideration. Accuracy is contemptuously flung aside and eloquence talks gaudy eminence over truth.

"I would rise from death's bed itself to welcome you," said a friend whose daytime rest we once disturbed.

"She was lovely, she was comely, she was fair. She had ptarmigan wings upon her. She was a bird of note. That is, she taught other birds to sing" – so said another Highlander describing his sweetheart.

You can take your choice if choice there be. But of this you can be sure, conversation in rural Scotland is still an art.

MY FRIENDS IN NEED (22)

For a variety of reasons we had to reconcile ourselves to doing major improvements to our byre. The old-fashioned causeys laid by the skilly hands of my dim predecessors had to be torn out and levels taken; stones had to be replaced, damp courses laid, and cement mixed and smoothed over all – no light undertaking anywhere at any time, but just now in the wilderness miles from a town with its rare workmen, well, I ask you!

We are getting on with the byre and this is how.

First the factor said, "Yes, I'll send you materials if you do most of the work yourselves."

Then the grieve said, "Yes, I've got notes of how to make byre improvements. If you get an extra Italian I think we could manage."

Then Frank said, "Yes, I've helped to do this sort of work before, and I know just the Italian for you, if you ask the Labour Officer."

Then the L.O. said, "Yes, I'll send the man you want."

And that was the finish of the first verse.

Somebody else then rang up to say they knew the very place for gravel for our cement. The man who does our carting came to cart and said – "I think you would be cheaper getting another firm to do this work for you. They specialise in carting and howking gravel."

In the meantime I had written a letter to a firm of tar distillers asking for special paint and good advice about asphalt for a damp course. Back came an instantly courteous and helpful reply telling me all I needed to know and enclosing the address of another firm who specialise in supplying materials for damp courses.

A letter to the second firm brought as kind and helpful an answer as the first one had done. Could they have particulars as to measurements and they'd do all they could to help, and they did not think I would need a permit.

So that is really the end of verse 2.

The next business was getting tools together. We possess many varied implements, but we did not have extra shovels. We did not have picks for cleaning out the old cement from the walls, and, finally and biggest lack of all, we did not have a big spirit level.

My near neighbours gave us shovels, the factor by various devious ways got us a pick. And that just left us the level to get.

For a moment we were in despair. Then as a forlorn hope I rang the garage which does our tractor repairs. Did they know where and how?

To my everlasting amazement and gratitude a crisp, efficient voice said "Oh, yes, I'll lend you one. You'll take care of it, won't you? I'm not sure where I'll get it, but you can depend on me finding one. Call for it to-morrow." So the man who does the carting called for it and there you are.

That is a true story of ordinary folk helping an ordinary body. In a world grown somewhat bleak I find it comforting. Don't you?

BY ANY OTHER NAME

16 September 1961

With the great autumn sales of sheep and cattle imminent we begin the annual decision on what beasts to sell, when and where. It saves much confusion if each beast has a name, a convenience which can be a tax on one's ingenuity. Finding names for 50 calves is wearing. Sometimes, of course, a heaven-sent inspiration suddenly names a beast Pearl or Shamus. But such happy thoughts are not on tap; and at one time I had to fall back on the Christian names in whatever classic I happened at the moment to be reading.

Thus we had Brutus and Manon, The Pirate and Clarissa. Then I began to suspect something odd. The beasts seemed to change their natures and start to act like the people for whom I had named them. Was it possible? Then I bought three heifers in a row and called them Charlotte, Emily and Anne. Emily and Anne died in early youth, of strange and lingering deficiency diseases which in those days we did not know how to cure. Being even poorer than usual, I fled to inspect Charlotte, who was looking even more red-headed and intense than ever and suffering from indigestion.

Hideous Calf

This sort of foolish coincidence is hardly worth mentioning – except that I never again called a beast any name that had not fortunate connections. Then last year one of our cows had a calf we named "Wee Hughie." He grew not in grace nor stature but he had a compelling character whose nature and power were instantly recognisable by beast and human.

The peccadilloes of any ordinary high-spirited calf were kids' stuff to Wee Hughie. His exploits bore the bold imaginative imprint of a superior mentality. No mere bashing of the electric fence to the ground assuaged his thirst for adventure. He preferred to lead the whole herd over the flood-ingested river and there above the roar of mighty waters smile at the blandishments of humans, including the vet, who waited to give an injection for tick to those who needed it.

In time we sold Wee Hughie and felt that a perilous chapter of farm history was ended. Some months later a neighbour came in to complain about how naughty her second from the end infant had become. From being a blue eyed cherub he'd become an obstinate little demon constantly leading others astray.

No Difference

"I've scolded him and thrashed him," said his angry mother, "but it makes no difference to Wee Hughie."

I staggered and falteringly repeated: - "Wee Hughie?"

To which she answered happily; - "Yes. After your calfie you know."

This was too much. I rested not till I persuaded her to get her man to look for another job – which he was thinking of doing anyway. Feverishly I thrust advertisements at her till she must have thought me mad. However, a new job was procured many miles away and this very morning Wee Hughie came to say good-bye. Clothed in white shirt of immaculate integrity he turned limpid blue eyes to mine and left for the beginning of a new life.

THE FIRST STOOKS

28 August 1945

The first stooks of our first peacetime harvest are up. Certainly, down the country they have been standing for a while, but it is only this past week that we have begun in earnest up here.

Seeing the stooks standing in regular lines like soldiers, we forget the awful agonies of anxiety which beset us before we started to cut. There is just a right moment to cut oats, or any other grain, for that matter. And the difficulty is to decide just when that moment occurs.

Ticklish Problem

An old farming saying avers that you should cut oats a week before you think they are ready, but barley is to be cut the week after you think it is ripe. But there is so much to be thought on. Certainly the oats at the top are ready to come, but down there in the lyth it is nearly as green as kale.

Of course it will ripen in the stook, but for all that what if there is no milk in these green ears? On the other hand the ripe stuff will shake and we'll lose the grain. Remember last year.

We finally decided to cut.

The binder has been ready these past two days. It has its new canvasses, its knotter has been given the once-over, the position of the flies adjusted, and the field is ready roaded, so let's go.

It is a poignant moment when the binder starts its first round. Here is the crown of your farming year. For this you have worked with your hands, your heart, and your brains. It is no wonder that you feel triumphant now. But this is usual significance, for one cannot forget that here is food, and that a whole continent, our continent, is starving. It is not selfishness that makes us pray for good weather and timeously gathered crops.

Not so Easy

As you watch the slow progress of the tractor with its attendant binder round the field you remember the war years. The incessant cry was for increased production and more and more skilled men were at the same time being called up.

There is a sunny and disarming belief among townsmen that any able-bodied person can do farm work. The farmer, poor wretch, knows that it takes about six unskilled folk to do the work of one who knows his job.

Building loads is not heaving up sheaves all anyhow on to a cart. Keep your corners square, heart her up plenty, keep your load well forward. How well the skilled man knows what to do. He is always at the builder's hand with directions to the right place. But it behoves you to be careful to criticise amateurs. Without their help we could not have succeeded as well as we did.

This year, however, the odd collection of children, students, old-age pensioners, and what-not is gone. Our skilled men are still in the Army, but our fellow-workers in the rich fields are Italians and Poles. Victors, defeated, humiliated and homeless, we all work together.

Tonight as we finished a perfect rainbow spanned the evening air. I wish I can think of it as an omen.

AUTUMN

EARLY AUTUMN

We can usually count on a gentle September in these hill parts and even a sunny October, but this year autumn came crowding in on the wet heels of a dirty August.

In such inclement times we are glad we managed to corral the last of the pullets and get them properly housed. After defying us with impunity from the tops of the larches and compelling us to the strangest stratagems, which varied from cutting branches to steeping corn in whisky, we found almost to our chagrin that all that was needed was a good stoning, and then taking advantage of their surprise, walking them quietly into another henhouse and so capturing them.

Tousled

Last week we had two good days and were lucky enough to get Bob and the Government binder then, so now we have most of our corn cut. The sheaves are gey tousled in places but, as we expected not to get them cut at all, we are pleasantly relieved.

No crop, however tangled, seems to upset Bob's large placidity, and he and his "Major" can reduce the wildest tangles to some kind or order.

Now we are busy stooking, and if we are not up to the eyes we can safely say we are up to the armpits. Stooking looks very easy till you start.

Anyone, you'd think, could put up those dear little children's houses, epitomes of a long-ago youth, when wigwams and Indian braves contended for our imaginations. Alas, that dreams should so quickly vanish in impatience.

Each sheaf is cut on the bias with a long toe of straw and a short heel which is grassy. Many people taking advantage of the bias place the sheaf with the toe to the outside.

But since the point about stooking is to get the stuff dry, it is better to get the grassy side out to the air.

Being short too, it will be off the ground and so all the easier dried. You will also have the dry corny head to the inside of your stook so that it will thus stay dry and everything will be lovely, more or less.

However you stook, it's a wet job in this weather, and nothing is quite so damping to the body and spirit as oxtering soaking sheaves.

Suddenly, though, the sun breaks through and reveals a radiant world for a moment where the scarlet leaves of the wild cherry sing out against the dull pewter of lowering cloud and wagtails dance a minuet in black and white on the purple roofs of the slated steading.

SILAGE "WI' NAE STINK ABOOT IT"

2 September 1961

We are bashing away as hard as we can to get our second silage pit filled. If it had been a normal year the wretched thing would have been done, but the long cold spring hindered growth so much that we began to feel we'd not get even one pit secure. Now we are delighted to have been wrong, and work away after suppertime till the great moon lifts in the autumn sky and the owls begin to hunt.

The last two days of warm weather have been a great help and comfort for they were accompanied by a round gold sun and a drying wind which still blows constantly in great good humour from the hills in the south. When it reaches the deep bonds of thickset conifer you can hear it rushing like a river through them. It spangs clean across the moor at such a rate that our visiting bees thumb a lift as it goes, and reach the heather in no time at all. For once it looks as if we are to have a real heather season with none of the washed-out effects so often our lot at the back end of the year.

Wide Eddies
The swallows too are delighted by the wind and they disport themselves in its wide eddies like children at the seaside. Many of the early birds have already gone. Probably the ones we see have been rash enough to embark on a second brood. There is always a great noise of cheaping in the rafters of the empty summer steading.

But in farming there is aye a something and the wind that is so welcome in lots of ways has the annoying habit of creating a certain chaos with the actual forage harvesting. The idea behind this operation is that the harvester engorges the grass, chews it up, and then directs it in a neat green fume into a trailer which is attached to the outfit. A strong wind catches this fume and blows it past the side of the trailer and leaves it lying uselessly on the ground. To avoid this takes much tricky handling on the part of the tractor driver, a task made even more difficult by the shape of our brae set fields.

However, it can be done, though the fields present a curious appearance as each cut gets more complicated, until the place looks like nothing so much as a hilarious exercise in geometry by a giant who could not get his problem out.

Awful Circus

Our neighbours, who regard our farming as a kind of awful circus, are waiting for their own orthodox harvest to ripen. Down in the Laich the combines are wolfing up whole fields of corn but as we are three weeks, at least, later the time lag does not worry. Besides, it gives our farming friends a chance to look in and shudder vicariously at the sight of good grass macerated and put in a pit as if it were tatties.

Our silage pit is conveniently sited to the road and this lets the passing pal draw up his tractor and tie it to the hitching post of an old fold. Thus safe from holiday traffic and the more pressing contingencies of shaken oats or knee-d barley our friend can brood in quiet and confidence. He supposes, he says, that silage must heat. Would we advise him, supposing he was to be so left of his senses, to give it a good "bile"?

Somebody else says that cold silage, that is, silage that does not ferment, just goes into manure – only manure is not the word he uses. Some one quotes an article he read about a machine that makes tattie silage with "nae stink aboot it."

Dod, who is spreading the loads and tramping them, insists that silage properly made has no off-putting smell. A cynic suggests he should send a bag of silage to the harvest festival at the kirk instead of a sheaf of corn or a bag of tatties. When he appears likely to take up the challenge I determined not to be present at the ordeal.

FROM THE BULL'S MOUTH

7 September 1948

I, Frankie-the-Bull, am speaking this. I am doing it because since the Human shut me off into a paddock half way up the hill road, I am bored to distraction.

Humans are poor creatures though, since I am a just man. I must confess they do their best for me. I get plenty of food and drink, for instance, but a man of my temperament needs more than merely bodily comforts. He needs sympathy, a broad warm outlook, sensitivity, in short.

In their oafish way the humans glimpse my yearning and do what they can to appease them. They realise I require company so they send me what they call company. Pah! My youngest daughters are sent up to talk to me through the barbed wire which fences me off from wider contacts. Oh, I can do with the children up to a point, but when "Papa, I have lost my hanky, give me yours. Papa, I want the clover beside your drinking trough. Papa, my tummy is sore. Papa, I'm going to be sick, papa – papa – " then I can stand no more.

I turn the ring up in my nose and I squeal. I have a lovely squeal, high and forked with malign intention, and when I scream "Eeeh-eeh" over my lolling tongue you should see what happens.

In no time at all up come my wives from the bottom of the second year's grass park all hot and bothered. "Franco how can you?" they pant with heaving bosoms and embrace their blubbering offspring. Good heavens, what did I ever see in such women? A man changes of course, but so do women – for the worse.

I don't recollect Bella having been so heavily domesticated when first I knew her. Of course, I took up with her as a refuge from Charlotte's tantrums.

There's a handsome piece for you, but what a temper! I believe the humans intend dehorning her some of these days and, though I'll be sorry since it will spoil her looks, yet my other wives will breathe more easily. Yes, she is rather torrid, is Charlotte, and quite different from Rhoda.

I can't think what ever attracted me to Rhoda unless it was her character. Goodness knows it could not have been her looks. Of course, she was one of those girls who go off early and now she is very plain indeed.

She doesn't help matters either by insisting on wearing cheap, utility shoes. Devilish slipshod. She never had good feet even as a girl. Still, as I say, she has character, which is more than I can say for some of my other fancies.

Heigho, I am bored, lord, I am bored. I wonder what like the girls on the hill are. Should I find one, even one simpatico? Perhaps? Oh, confound this barbed wire.

Signed,

FRANKIE-THE-BULL,

per Elizabeth Macpherson

NIGHT ABOVE THE VALLEY

11 September 1945

The prosaic job of emptying my dungaree pockets of corn took me to the back door to-night. Because my work˙was done and because the night was bonnie I lingered to enjoy it.

Night mist and vapours crept up from the heart of the valley, and as they came they brought with them the unmistakable elegiac scents of autumn. The heather is already waning, but it can still fill the air with its honey sweetness, and the first hint of frost had brushed the rough leaves of the turnips so that their harsh aroma mingled with the scent of the heather.

The fields are all cut now, and where a week ago the corn waved shock-headed stooks stand in orderly rows facing north and south. The last reluctant rays of the sun caught them and gave their gold a ruddy tinge, and then gentle dusk came to wipe away their colour and leave only their shape to tell that it was still not night.

Idly I wondered how long it would take to lead the crop. There are so many ifs. If this weather keeps, if there are no stoppages, if we get extra help, and finally, if the stacks keep.

We do not grow any grain but oats here, so we are spared the acute discomforts of handling barley. Oh these vicious barley yavins – awns, if you like it spelt that way" They get into your clothes down your neck, and into your eyes.

Oats are pleasant to handle, especially the ones we grow here, for they have a good strong straw which makes them very easy to build into loads and stacks. Most of my neighbours grow what is called a "tattie" oat, which has a fine sweet, good feeding straw, but in bad weather has a tendency to go down and so make difficulties in harvesting I prefer to play safe and grow stuff called Yielder, and my crops stand.

As I stood contemplating, the sleepy noises of night came towards me. Some one far away was driving beasts with a dog, and the commands came echoing across the distance. There is magic in the disembodied voices. They have a pure and bell-like quality, like something remembered or heard in a dream.

Lights now began to ring out in all the little houses. We never experienced here the rigours of the town black-out, but the friendly lights of houses inhabited by men were a sad lack in the dark. Tarbat Lighthouse shines out in mercy over the shining water beyond the dim woods.

At last it is fully dark. Above me swings the Plough to remind me that after harvest we must prepare for next year. For a moment one feels on the brink of a revelation. But eternity and immortality are too great for a poor human to apprehend and the moment is gone. All we know is that somewhere there is music in spheres.

TOO GOOD FOR THE STACKS

18 September 1945

Like everyone else here, we have been leading in the harvest all this last week. As golden day succeeded golden day like notes in a song, the big wagons lurched and swung up through the brae set bottom field and home to the stackyard.

This year we decided to change the position of the stackyard from behind the steading to a piece of hard ground nearer the road. This, we hope, will not only give us better bottoming but also an easier ingress for the threshing mill. Last year's adventures on a glaury winter night with wind, rain, lanterns and an infuriated Highlander are still too vivid for us to wish a repetition.

The weather has been unbelievably beautiful. The sky is blue with clouds so carefully arranged in broad layers one might think one was looking on a painted scene. The sun drips gold like a Tennyson lyric and the soft airs barely stir, so heavy laden are they with the scent of honey and ripening apples. The bees, drowsy with too much nectar, labour home, and from their crammed hives comes the perfume of matured honey.

Unfortunately this exquisite autumn is not quite what we want at this time. In order to winnow the stooks thoroughly you require wind rather than sun. And for once wind is just what we do not have.

The vast amphitheatre of the sky is calm and dreaming as if it had never been the arena where Boreas sported and played. Each night the grieve whistles for a wind, which I tell him is tempting Providence, for rain is wind's companion. But so far the heavens have paid no heed to him, but wheel serenely on their astral business impervious to the mortal supplicating for a good drying wind "before the stacks settle."

The new stackyard is almost full, and by to-night it will be finished. The stacks themselves look snug and comfortable. I admire their shape, at once trim and rough, like well made suits of Harris tweed.

But the grieve is proof against my admiration and says darkly, "Handsome is as handsome keeps. The bonny stack is not aye the best keeper. Forbye, this weather is right bad for stacks."

I feel his attitude is somewhat ungrateful, though I know full well that he speaks truly.

Yesterday we all got the wind up because the man with the petrol lorry who was delivering our paraffin announced that people down the country were turning their stacks because they were heating. They had been lured by siren weather to take home their corn before it was properly cured, and here was the result.

We hurriedly thrust arms into our stacks, and were thankful that there was no sign of warmth about them, and further they were rocking as all good stacks should. We shall be relieved, more than glad, when all is safely home. When we can look at full stackyards then we can say that once more we have triumphed, but until then we are keeping our fingers crossed.

JUST ODD THINGS

28 September 1948

Everyone here is harvesting madly, the men keeping one eye on the irascible weather and the other on their harvest volunteers, what time their sorely tried wives try to part out non-existent harvest rations over an extended period of time.

I never could see the point of post-harvest junketing, the prevailing feeling of myself and my neighbours being one of wan relief that the whole worrying time is safely past.

This year the business has been more exacting than ever; although we all assure one another that we are luckier than the boys down the country who were nearly driven to the asylum by their swirled and flattened crops. Our crops for the most part stood fairly well, but the curious weather had the odd effect of ripening them all throughither so that we, for example, have finished leading the ley while we still have a whack of clean land stuff to cut.

Still the weather at last looks like settling up. It is dry anyway, though so cold your nose is snapped off the minute you put it outside the door of the morning.

To put a spur to our intent too, the hills of Ross are wearing the first whitening of the year.

Frost ends the short day and sharpens the brief twilight with its bright pang. Trees are turning, grouse are packing, and the robin pipes his sweet autumnal note within the gean.

You feel the cows should now be coming in at night, the worry of feeding-stuffs looms (for, like a lot more people's our neeps have felt the bad effects of the early, incessant rain), and finally you awake to the fact that you haven't yet filled in your September agricultural returns. So you try to do it while turning a batch of pancakes on the girdle. Take it from me, this is a very, very difficult thing to do.

Altogether you feel all worried and pushed. If only you had even one half-hour to think in.

Then you could arrange for the milk cows to go to another field, since they are tired of the one they are in.

The calves, too, are needing redding up, for Vitella is not going to bring her last one through, so it will have to be switched to Vaness, who'll protest something horrid.

The grass is going back, so the whole lot of milk cows will require extra hand feeding.

The corn will soon be in, however, and then we start the tatties. What jolly fun!

LOVESOME PLOT

30 September 1961

One of the immemorial laws in agriculture is that the farm garden is the province of the farmer's wife. The idea behind this comfortable convention is that the farmer himself is so exercised with big important policies involving thousands of acres that he cannot bring his mighty mind down to the level of two grass plots and a flower border.

Like others before me I meekly accepted this convenient masculine ruling and have coped away, not without some pleasure, with a garden whose origin was an old peat stack and whose boundaries are visited by all the winds of Dava. There were indeed times when I felt frustrated, and then I'd lament my pitiful state to my husband. Then he would make helpful gestures, not so much for my sake as because he cannot bear to think of the earth not doing its duty and growing things.

Garden Fence

The trouble was that his gestures were so lavish. I did not really need 10cwt. of ground limestone dumped over the garden fence when an old sheep-dip pailful was all that was necessary – forby all that I would carry. There was too the matter of farmyard manure. This was brought to the gate by tractor and cart. Dod did say that it was a pity I could not have got the muckspreader into the garden as it would have done the job in no time. One skelp from the thing would have knocked the house as well as me and the garden flying. So I attacked the malodorous heap with the kitchen shovel and my hand pail.

Such fatuity was too much for Dod who ordered his current helot to graip the stuff into heaps convenient for me. I just had not the nerve to say the heaps were just as difficult to handle since I totally lacked the technique for sending them out into nice even circles. So I just waited till he was good and far away, at the mart or at the bottom of the hill drain, before creeping out with a wee trowel and the afore-mentioned pail and generally using my own method.

Such experiences teach one to be careful what one says. But alas there is always, in my case, the frivolous thought, the foolish utterance. That was why I said how delightful it would be to grow strawberries. I might have known Dod would take this at the foot of the letter.

He departed to the telephone where I, all unsuspicioüs, imagined he was making arrangements for a float to take beasts to market. What he was doing was ringing up appropriate departments for useful knowledge on how to grow strawberries. In a day or two I was inundated with pamphlets and booklets on strawberry growing. Whiles, strange men rang me up with more useful knowledge.

What peculiar diseases such luscious fruit is heir to! They have peculiar domestic habits too and you must plant them good and far apart so as to make them keep their evil influences at home. You must be ruthless with their anxiety to propagate themselves by sucker – I thought this quite revolting – and finally you have to nip them in the bud. This seemed to me to be forever putting off the time when you might expect a berry.

New Plants
Some one else said you had better divide your garden into three so as to have room for new plants to grow up to take the place of their grandparents who were then to be consigned to the flames. However I duly took in all the instruction and hoped to put it to use, though for the dozen odd runners I meant to grow I thought the information a little overwhelming.

At last when Dod thought I was sufficiently educated he brought home the plants. There were 200 of them.

Probably farmers' wives would fare better doing the garden themselves and keeping very quiet about it.

HUMAN MOLES ARRIVE BY BUS

23 September 1961

Now that the silage is off our fields the pylon boys have moved in with such a raft of heavy equipment you'd imagine they meant to bulldoze an atomic site out of our hill masses of granite and gneiss instead of just digging a few mole holes for the bases of the pylons which are to swing their powerful airy way from Beauly to Kintore.

The men who come to work the robots poised so precariously on the cliff above the river arrive by bus. During the day this battle-scarred vehicle reposes her panting flanks in one of our cattle courts announcing between sighs that her improbable name is Eve.

Because the year is advancing and the first frosts of autumn already flickering in the high thin air, there is a feeling or urgency about the mole holes since the contractors do not wish to be caught in one of our notable Dava freezes which petrify the very bowels of the bog. Therefore huge piles of rubble accumulate with speed and precision along the lip of the crag, and the mesmerised spectator cannot cease marvelling at the contract between the suave co-ordination of the man-made muscles of the great excavators and their grunting, wrenching shoulders working their way through rock and bank.

Sometimes the workmen come across a band of stone, and their machines are slowed up. Then they blast the impediment asunder. On a still autumn day you hear the abrupt crack of the explosion and then a moment after the silence heals itself in great concentric circles of ameliorating echo.

Naturally such impressive noises cause loud comment and endless conjecture among the people in the glen. Some attribute them to use. After all we have no visible harvest and yet manage to appear solvent, so probably we are howking our wealth out of the ground. The blasting on the hillside is but another aspect of our peculiar farming policy.

Now one of the drawbacks of this glen is that there is no place where we can meet one another informally. In the village hall we lead entertainment lives exclusive of the other sex. So the only meeting place common to us all is the kirk. Naturally our puritan history rather inhibits us from free and chatty inquiry about one another's doings in such environs.

But human nature will out, and it is wonderful what you can do if the organist plays a psalm slowly enough.

"Oh thou my soul bless God the Lord" – and while you wait for the next phrase you assure your neighbour that you are not building another silage pit. The hydro boys are at it again. "And all that in me is be stirred up" seems a natural as we both contemplate with annoyance the bother that such installations will cause the poor farmer who will have to cultivate about the stay ropes and huge concrete bases.

Perhaps a little guiltily we concentrate on verse two. Our souls are after all too fond of being forgetful of all the gracious benefits that are bestowed on us. I look out of the little old windows and see the old-fashioned rose blowing against the gravestone erected to the memory of Christina Grant, who died these many years ago on one of the more outlandish of our hill farms.

Her descendants are here today; they came by car and know that when they go home their dinner will be ready in the electric cooker with the time-switch in operation. Christina fetched water from the well and cooked with a peat fire. I expect she'd have thought electricity a very "gracious benefit" indeed.

"All thine iniquities who doth most graciously forgive." Och well David was a kindly man and won't count discussion of the mole holes against us. We sing piously to the end of the allotted verses.

HARVEST 1954

We began cutting the ley oats to-day in the teeth of a September gale. The blue background of the autumn sky is filled with flustered clouds whisking their petticoats away from the south-west wind and obscuring the angry sun, so that he sends his light in great yellow gouts whose only merit – albeit a fictitious one – is that seen thus the corn looks much riper than it actually is.

We know perfectly well that the stuff is on the green side, but we are three weeks behind, and of what October and its storms may bring we do not like to think.

If we wait till the corn be ripe we may have no corn to cut at all; so that the smaller loss seems more endurable than the greater risk.

I never remember a year when the corn was so sweirt to turn. Even the barley seems to be making more of it than the oats this year. The truth is that we have not yet exhausted the legacy of our bitter summer.

The wild haws and the rowans remain an obstinate jaundiced green, and the local apples of the Laich, usually so sweet and juicy, taste more like the grapes of wrath than apples.

It is true that down in the Laich things do look a great deal more kindly to the hill farmer. Yesterday everybody was in the harvest field on the outskirts of Forres.

The rule nowadays seems to be that you either have a combine harvester, which processes the grain direct from the field to the grain store, or you take the mill out into the field and thresh direct from the stook to the sack. Either way, you get out of the laborious business of leading and stacking.

I suppose what you save on the labour bill for the manual labour which this entails more than off-sets the loss in the grain cheque brought about by the high moisture content of the corn.

I'm afraid, though, such rationalised methods will not work here. Neither our climate nor our hilly fields nor our bank accounts will support the techniques suitable for the big Laich farms.

Anyway it is only by the grace of the marginal grants for fertiliser and the brute strength of the individual farmer that we grow corn at all here.

Gradually we'll become again what nature intended, live stock rearers.

We'll always do some arable farming, of course, for the sake of our beasts, so the stack – that ancient symbol of security and plenty – will always be found among the hills.

LIGHTS OUT

1956

Everyone is by now familiar with the outlines of the litter system of keeping hens. By shutting the silly creatures in a warm house with lots of artificial light you make them think it is springtime. By judicious juggling with the daylight you arrange to give your hens 14 hours of light, but however you juggle you must always do your trick at the same time.

Where farms have electricity, of course, this presents no difficulty, as all that is needed is a time-switch. Unfortunately we have not yet reached this happy bourne of convenience and must make do this winter with bottled gas, which, though good, still needs the human agency to turn it on and out.

Lighting-up time is very easy because I am down at the litter house, which is at the other farm, doing the chores anyway. As I come home by a short cut through the park I like to rest the heavy basket of eggs on the ground and look back to see the lozenges of light from the skylights growing brighter as the swift dusk falls.

Cold on the Feet

I wish Lights Out were as easy. We take it in turns to do this, and 10 o'clock at night sees one or other of us prowling down the road with a torch, a brave but pathetic illumination against the wastes of Dava. Leaving the warm fire and comfortable slippers and climbing into wellingtons may be good for the character, but it's awful cold on the feet.

On dry nights the cats accompany me weaving in and out of the light as soundless as their own shadows. They purr, however, now and again to let me know I do not walk with ghosts.

On moonlit nights the world is bright and exaggerated, but when there is no moon only the scattered lights in the valley assert that men still live in the desolate dark.

The litter house itself is warm and domestic. While I collect clockers and renew water the cats steal into the bin where the pellets are kept and eat like greedy burglars.

The pellets are strongly reminiscent of fish and tactlessly remind me that it was not affection alone that brought the cats. When the night is wet I have no company. The road, black and shining like a whale's back, swoops and surges through the ocean of the night. Long after I have left it it will go on, Leviathan swimming over into Strathspey.

TIME, PLACE, LOVED ONE

We are not, as might be supposed from the title, engaged in a factious romance. Instead, we are trying to make arrangements for our tattie lifting, and the only constant in our unmanageable trinity is the place.

This year the potatoes are in the furthest away field in the place, and since last year they were at the back door we have all the duffle of straw, scows, bags, pails and tattie digger to transport over hill and dale to the tryst. The grieve attends to the sordid details of digging pits and refurbishing implements, while the telephone and I try vainly to make the time and the loved one come to an agreement.

The real trouble is that the loved one is not one but twain, and consists of tattie lifters on one hand and a tractor man with a tractor and a bogie on the other, and unfortunately these irreconcilables are under different directors.

There are other complications too. In this sparsely populated countryside there are not enough children to be had, so we, still under war-time compulsory cropping, must seek our casual labour outside our own bounds.

The late harvest has meant that our glossy brethren of the plain are still busy in their vast potato acreages, and since we are only little wee noises we must wait till our betters are served.

A harried labour officer does his noble best, but he cannot promise either a full squad or a definite date. All he can say is that I can be sure of lifters in the imminent future.

OK. Now for the tractor man.

A matey voice at the other end of the wire says breezily that Jock the tractor man with a tail of implements is somewhere in the fastnesses of Dava Moor and maybe I can go and dig him out.

"Fix it with Jock," the voice continues and since "fixing it with Jock" is always an entertaining business (for Jock as well as being the soul of helpfulness is full of the more cheerful affairs of our scattered neighbourhood) I depart by the light of the moon to contact the gentleman.

"Och, ay, I'll fairly come," beams Jock, "but I've a week's cutting on my hands yet."

So here we are with our irreconcilables still irreconcilable and this St Martin's summer slipping past. Time, place, loved one? There was never a love affair half as temperamental as this.

"THE RAIN IT RAINETH EVERY DAY"

For a long time now we have watched wet days lengthen into wet weeks and dolorous summer hand her willows to an equally lachrymose autumn. Vainly we try to comfort ourselves by saying that maybe our lateness will yet save us, for we have nothing ready to cut so far, although the thin stuff down on the poor land could come mostly any time if we got a day or two's sun.

It is not a pleasant feeling watching a whole year's work threatening to topple into ruins. We occupy ourselves sickly, tidying the barn, mending bags, rolling up Glasgow Jock (rope) to secure our stacks (if ever we have any), but our hearts are not where our fingers are but up with the ley corn, which is a heavy crop, and heaven knows how much longer it will stand the torrential weather.

At night I lie quaking, listening to the angry feet of thunder showers and think with sympathy of my neighbour whose lovely crop of oats was battered as flat as if it had been steam-rolled in last Sunday's deluge.

The awful thing is that you can do nothing about it. No piety, no wit can avail, and you must just thole as best you can.

Down in the Leich, the big farming barons are busy in every blink with combine harvesters which cut and thresh the crop in one operation, but I have my doubts as to the condition of the grain thus garnered. Of course, it will go straight to the hungry mills and get dried there, but it would be a miracle if with this mucky weather the machinery could shake all the grain from the straw.

However, desperation makes wastrels of us all and half a bag of oatmeal is better than none.

In books and in pictures harvest looks very nice; autumn drowsing on the granary floor, her "hair soft lifted by the winnowing winds" is a lovely mellow thought, but alas in a bad season you don't have thoughts like this. Anger and impotence lay waste your heart, and you recollect old tales of harvest still lying tattered in the withered fields of bleak snow-whitened December.

Yet I know that, given a single week of sun, we'll forget all the agonies and anxieties and disbelieve that rain ever fell or furious winds raged, so brief a memory have we for misery, so tenacious a hold of happiness, accomplishment, and the sun.

HALF DONE

1 October 1946

I am going about just now in such a fither of irritation and anxiety. My soul's feet feel as if they were walking one in the gutter and one on the pavement. It must be abominable to have to live with me. All for why? Because we are still harvesting.

We have had such a nice time getting the stuff cut. There were great lying holes which had to be scythed out, and the corn thus cut had all to be gathered by hand and bound. So primeval, and so slow.

The field with the heaviest crop was a brute to cut. Talk about heihs and hows!

And where it was not braeset it was adorned with great boulders which were young during the Ice Age and which now are in their prime. The tractor went temperamental. The binder went temperantaller, the grieve retired like angry Jove behind a thundercloud, only it seemed a handy cloud, because he emerged from it complete with a set of new sparking plugs and a fancy fixing for a binder canvas.

With all this vexation I've hardly noticed that autumn is slipping by unnoticed and unenjoyed.

Do I care if the sunlight lies like silk along the old drove road, suddenly visible amidst the tawny orange of the moorland grass? I am quite careless of the blue metallic sky where the first frosts faintly chime. All the winds in the world can chase the argent galleons of September down the sky and I shall not see.

All I can think or dream of is when the sheaves will become stacks and when the swept fields will await the winter. At night I go to call the cattle home, and most unseemly bawl at them to hurry, for all that is not urgent and scourging seems vain and tiresome.

Rowans, their orange berries shrill as elfin calling, are as nothing to my hurry, and the humble vagabonding briar, for once kingly in purple and cardinal red, is only an interruption to my flying scarf.

They speak about harvest homes. Gracious barns hung with the gold of gathered corn and lit with kindly lamps, of empty floors waiting the dancers' feet, of groaning tables laden with food and drink for the exuberant, triumphant rustic swains and their "country girls with faces round and smooth as pearls".

O.K. You have it.

When this hairst is over I'm going to bed for a week. I bet my family and staff will welcome the ensuing peace.

BLUSTERY OCTOBER

5 October 1948

October came in on the wings of a terrific gale, which although it has the merit of winnowing the stooks still remaining in the emptying fields has the serious disadvantage of making these same stooks most unmanageable when it comes to getting them home.

It's no fun forking up sheaves to a cart in the middle of a hurricane and its no better fun for the poor wretch aloft who is trying to build a load.

Building loads under calm weather is tricky enough with the leaves coming pell-mell, but when they come propelled by a boisterous wind hallooing along at 60 miles an hour then you could cry with frustration and wonder why the so and so you don't sell out and go to sea.

Yet it is pleasant to be outside and part of all the blue, clean, windy world. Black showers loom out of Dava and go sulking along the spurs of the hills to vent their spleen down the country what time they leave us in peace to get on with our work.

The trees are changing, but with all these gales we shan't have a chance this year to enjoy their colouring since their gold and scarlet is almost at once whirled off and their poor bare arms are left to catch all the rheumatics of the day. But the leaves are not the only casualties.

The rowans, which were very plentiful, are now torn off and strew the road while the birds lament the waste of their prodigal harvest.

On the hill the tough short grasses are losing their freshness and daily the little hills become tawnier. The heather is over too.

They took away the visiting bees last night, the lorry careening wildly as it lurched down our rough hill road, all laden with bee boxes. The owners are glum enough about their heather honey results and all most of them hope for is that they shan't have to feed their bees the entire winter through.

Honey, of course, is no longer the fantastic price it was since housewives look twice at their sixpences before blueing it on a single pound of heather honey.

Mornings are now darker than they were, though I am still reluctant to take the Tilley to the byre to help me forage in the innards of the big zinc bins for calf nuts.

I wish the neeps looked more bountiful, for then I could afford to ignore the awful warnings in the grieve's voice which comes to me above banging floors and wind-rushed straw, "Ca canny with the feeding stuffs."

As we came from the byre the sun rises with a "View Halloo" from its eastern hills.

A WINTRY SNAP

7 October 1947

Snow has suddenly covered the hills to our west and north with a most unseasonably early mantle.

Our views thereby have been sharpened into shining and celestial loveliness. Unhappily though, we are not in the mood for enjoying the sight of the glistering ramparts of a fabled land.

No, we are farmers and we are still in the middle of our autumn jobs. We still have potatoes to lift and threshing mills to attend and for both we want moderately decent weather.

Wherefor when icy showers and blustery winds assail us every hour we shake our disappointed heads and, disregarding the penitent sun painting frail rainbows across the angry walls of heaven, we retire to gloom out fresh plans for our work.

Potatoes and mills are the special headaches of the hill farmer, for more than any other they depend on having people and there aren't anything like enough of them here.

Nothing exasperates me more than to hear townfolk being gently melancholy about the emptying glens, especially as all the time they are shedding their tears they are busy seeing how much more they can empty them.

They used to speak about evictions in the old days, but now they call it organisation.

Look at the place where we live. We send our young folk by bus off to Forres School 10 miles away. Forres is our nearest town. I am not clear in my mind why we do this except to encourage the little dears into training for living in a town.

When we are sick we must send for a doctor to this same town and unless we are nearly dead we feel most apologetic about asking him to come all this way.

Should we require the district nurse we must again send to town for her, realising the while very humbly that our concerns of aches and pains cannot mean very much to some one who is a complete stranger to us. After all she is human and her interests must be among the folk she sees every day and not with aborigines like us.

Happily we are law-abiding, which is just as well for the nearest policeman is also down in the promised land of the town.

Mind you, it is good that so many things are to be had in Forres. Otherwise we'd have to go to Elgin 26 miles away. I have to go there anyway if I want to send a parcel to devastated Europe or change a library book.

You have to be tough or stupid or thrawn to bide in the country. Preferably all three.

SAFE HOME AT LAST

8 October 1946

Well, that's that. As I write this the men are putting the last sheaf on the last stack and that's the harvest home. According to the old country saying we have now gotten winter.

For the hundredth time I reflect with gratitude on the luck which attended us this grievous year. Not once since we started to lead has the smallest shower of rain stopped the work, and we have been able to take the stooks right ahead instead of having to wile among them for the driest.

Now that it is all over we feel a bit light-headed, but each entertains his own private satisfaction. Geordie, the cattley, reflects that now he can resume his cadet drills of late so interrupted by after-supper work. For we have had to make use of every single moment, and one night actually had to light a lantern to see what we were doing.

The grieve walks slowly round his stacks admiring them and wondering if those of his own building are as good as the ones his father built for us. He is promising himself a jaunt to the town, for he has not had a half-day far less a weekend since I don't know when.

In the house Mary and I are holding a mutual admiration society, for I think the way she has spun out the harvest jam ration is little short of miraculous, and she thinks the way the understanding and obliging baker and myself have managed to salve a few of our extra BU's is equally deserving of praise.

The very cows and hens take part in the general glow, for now the empty fields lie open to them to be loitered through and picked over. Indeed, the only discontented member of our community is Charlotte, the big red cow, who took pneumonia in the middle of all the stramash.

We shot a pound of M and B into her and tied her into a sack waistcoat, so she is convalescent, but confined to the byre still. As she sees her companions make their luxuriously expectant way to the ungleaned fields she bawls her vain vexation.

Tomorrow I must see what labour I can scran for the tattie lifting, but tonight I can go for a walk round our small kingdom, secure from worry and happy to enjoy the sight of stubble fields fulfilled and peaceful stretching pale down to the quiet burn.

DOWN IN THE HOWE

14 October 1947

For reasons of space we must outwinter the heifers this year, the yearlings anyway. Many people down the country never do anything else, but we are so fearfully assailed by winter's "battering siege of wreckful days" that hitherto we have not had the courage to make the experiment. However, needs must when the byre is crammed.

So this afternoon I went out to see where we'd best put the beasts. I think I have found a suitable place down in the howe where the river runs.

The fields which edge the howe always seem to me to wear an air of comical surprise, as if they'd suddenly stepped on a banana skin. This open-mouthed appearance is understandable enough, for the ground at their far side quite on an impulse falls sharply away into a steep declivity some hundreds of feet in depth.

A mad zig-zag path flying headlong down lands the traveller breathless and a bit alarmed on the valley floor. Yes, there is a fairly wide valley here, and through its level centre runs the river telling a tale of green days to the smooth stones.

An air of deepest peace permeates this secret place. Birches, thick in their last god, line the sides of the precipice, and although you know that you have left a gale up aloft there is no sound of it here save where you can see the thick foliage moving and swaying.

On the valley floor there are thickets of arn, called by country folk Scotch mahogany. There are hazel copses, too, and nuts for the squirrels.

Sometimes a stone dislodges itself from the sandy craig and patters its helpless way down to bury itself in the funny green hummocks of moss which rear over every fallen branch and tree root.

In these fuel-minded days I regard all this harvest of irreclaimable firewood with frustrated eyes. Never was the proverb "So near and yet so far" so bitterly apposite, for I don't see how on earth we are to get these logs up the difficult path.

Round past where the craigs make a blunt promontory there is a rough meadow which owing to its sheltered position is still full of good, natural feeding; for the beasts need food as well as shelter. There is still more good feeding along the river banks, and I think the cattle ought to last out here till Christmas.

Today the howe looks peaceful enough, but I have no guarantee it will always be so. Och well, we'll have to take a chance on that.

SCAPEGOATS WITH SPLIT PERSONALITIES

14 October 1961

One of the minor comic hazards of country life in sparsely populated Scotland is that one must accept the fact of the split personality in otherwise normal people. "One man in his time plays many parts" says Jaques of uncomplicated life in the Forest of Arden in Elizabethan England. We go much further in rural Scotland, for we not only play many parts but we also play them all at the same time.

The other night a haggard badminton secretary rang me up to ask if I'd come and chairman a concert his club were holding for the benefit of their funds. Willing as I am to oblige, I like to make a few cautious inquiries before committing myself to rushing in where other people's toes may be. I was well aware that in this instance the club lacked not members who were neither inarticulate nor blate, so how come this avoidance of the public chore?

Top Brass

It turned out that though they were indeed members of the badminton club they were but members; in another club they were office-bearers and exalted personages, in fact Top Brass. Now this second club had decided to hold a money raising function as well and as ill luck would have it had chosen the very date the badmintoners had chosen. We have but one hall and the rule says, first booked first let. Badminton had been first so the second club had to retire – which it did with great umbrage and such annoyance that its office-bearers refused to co-operate towards the success of a do which would have been of service to them.

No one would as much as sell a programme or even operate the lemonade uncorker at the dance which was to follow the concert. Mr Harold Smith who doubles the office of Children's Officer in Staffordshire and Burton-on-Trent has nothing on us. True we do not write ourselves letters. We just cut ourselves dead, which when you think of it, needs a deal of subtle ingenuity.

The habit of the split personality can lead to some peculiar situations. Think of what will happen when you are at once the hall secretary, a lowly scene shifter in the dramatic, and the big cheese in the rural. You are asked in two letters, which arrive by our solitary country delivery, for the use of the hall for the dramatic and the rural on the same night. Do you take the matter to prayerful consideration, spin a penny, or just give in to natural vanity and plump for the rural?

Blood Relations

No doubt the outsider may criticise an arrangement whereby one individual wears a triune crown, but what are we to do? We just have not enough persons to allow us the extravagance of one to each office. Added to which is the fact that the folk who do live here are less a population than a tribe, since everyone is related to the other.

This blood tie is like love – many waters cannot quench it; forbye it has its uses. No one belonging to the tribe lacks friends or help in his domestic sphere. But when wider issues are in question the tribal taboos move majestically into frustrating position and we are stuck. This is where the outsider has a vital part to play. Hitherto he has been leading his own lonely neutral life but if the communal wheels are to turn at all he must take over. He must realise, however, that such behaviour demands from him, too, the sacrifice of the split personality. If he naturally and easily assumes the role of leader he simultaneously becomes the scapegoat also.

I often think the essential benefit the country dominie and parson bestowed on the places where they lived was not the number of scholars they encouraged nor the souls they saved but the wider horizons they made all their people glimpse and their success in fusing the leader's and the scapegoat's position. It is from this they derived their deserved prestige.

TATTIES NOW

19 October 1948

We had hardly got in the last sheaf and got the hens on to the stubbles than the potatoes were on us – ten acres of them. Harvesting tatties is just as worrying as harvesting corn, but for different reasons.

In the corn harvest a shower of rain is enough to stop work for a considerable time, but an odd shower or two does not stop tattie lifting. No, the headache in the potato harvest is labour.

We are never more conscious of the troubles of an emptying countryside than at this time. Even hands for the mill are easier to get, for we can manage that somehow between us. But, oh me, the tatties.

Down the country where there is a pool of labour available in the adjacent towns things are not so bad. Or at least they used to be easier, but this year everyone who could increased their acreage – a legacy from last year's panic – with the result that all school labour is booked and more than booked for the Laich.

Then p.o.w. labour is at last no more, so it begins to look that if next year you want tatties you'll have to grow them yourselves.

We, marooned in our heights, are in a difficult position. The school children are already booked, so what can we do?

We seize our handy little telephones and we bawl through them for the Labour Officer. Maybe being a farmer isn't always fun, but I would not be a Labour Officer for a herd of attested pedigreed Ayrshires or a hundred tons of stock seed Majestic.

We are simple folk we hill farmers. When we want anything we yell, convinced like healthy children that the louder we bellow the quicker some one will come and give us a sweetie.

And we expect the Labour Officer to be that some one. What an edifying time he must have. I wonder how often he explains that two into one you cannot.

And then, too, he has the entertaining spectacle of all the Big Farmers outbidding each other for the available school children. Up here we too shake our heads over the unsavoury sight of fellow-farmers cutting one another's throats. We are not big enough to do that – yet.

However, eventually our gang arrives, and in the intervals of the ensuing labour we take time off to admire the extraordinarily effective blend of inevitability and world weariness which distinguishes the Labour Officer's telephone voice. He's a One.

But, believe me, I'll have not a quarter of these tatties next autumn. I'm telling you.

GEESE GOING OVER

21 October 1961

The great autumn migration of geese which began last month is now in full flight. Every day we see the great phalanxes and skeins tangling the mild October sky and hear the birds' plangent cries as they row their powerful way up from the firth over the hills to the south. Occasionally at night we hear them as they sail down the streaming stars and then their loud utterance assumes a wild and raucous majesty. Lords at once of air and water they but emphasise our poor earthbound lot.

When I lived near the coast we used to see huge flocks of geese feeding in the harvest fields and we called them corn geese. By the time they arrived the corn had been already cut so no one grudged them their gleaning. Perhaps, too, we were the more hospitable since we knew they had no intention of remaining. In two days they'd have gone.

In the hills it is different. Only once in more than 20 years have we seen them thus feeding. They came in at sunset and all had made landfall by the time the moon rose.

Like the hardened travellers they were they posted sentries before settling down and when morning came all had left save a few stragglers who looked mildly at the collie nosing in the dewy stubbles. They seemed quite imperturbable, and when they were quite ready embarked once more with dignity and care upon their sky-ey way.

This was the only occasion on which I was able to observe them close at hand here until the other day. I was going up the narrow defile which leads from the farm to the hill on a silver placid morning when one still mood enveloped firth and mountain. The collie breenging in the rape field put up a pheasant who hastily made for the thick plantation which makes a sheltering wall for this track. As he banged clumsily about trying to hide his gorgeous discomfiture in the sweet smelling wood the geese came up overhead as near as every I have beheld. Black against the lucent sky, they held their arrow perfectly and as they went they kept up a continual liquid murmur among themselves, a domestic gossiping quite unlike the harsh sound with which we usually associate them. The velvet precise beat of their wings filled the narrow gorge where I stood gazing and transported.

Echoing Call

As they reached the Knock they broke formation and a few birds seemed confused. At once the echoing call broke out and the wayward resumed their places. This is not always so. Sometimes I have seen birds lose contact until forming their own skein and choosing their own leader they have made back upon their track. Perhaps they joined up later with another fleet.

Occasionally we have wild swans going over. Like other amphibious birds such as ducks and geese they, too, favour the arrow formation, but they never crowd together like geese. Five or six birds make up the swan flights here, though you may see them in scores swimming together in the sea water at Inverness. Swans are not garrulous like geese and when they fly the noise you hear is generally the powerful rustle of their great wings creaking like wind-filled sails.

Here we associate the flight of swans inland with severe weather. Whether this is superstition or not I cannot say, but I do know that in prolonged wintry weather they come inland and take up quarters on isolated Highland lochs. Clustered on the dark bosom of a forgotten loch at dusk their whiteness assumes a luminous quality so that they make a vision of beauty that is unforgettable. They call out as they rise from the water and trumpets answer from the crowding hills.

THE NON-GOLD STANDARD

22 October 1946

We went to visit a neighbour the other night to see if we could get some hens' corn. We have not yet threshed and we had heard by various country bush telegraphs that he had. But the corn was not the sole reason for our call.

I really wanted to know why, if my neighbour had had the mill, he hadn't asked me for help. He laughed and said "Oh, I knew everyone was busy leading and I hadna the face to ask for a hand. But I'll need a hand again." I was much mollified.

One of the things which makes life in these Highland farms so pleasant is our tacit acceptance of the fact that we are all dependent on one another.

I borrow a horse and a broadcast from one friend and she knows that she will get my tattie dresser in exchange. I give hospitality to another friend's bees and he gives me a great basket of apples in return.

There is a constant ebb and flow of obligement and kindness, and the question of money never enters.

Of course when one accepts this system of exchange one also accepts its implications, which are quite onerous. You cannot do an obligement to a man if you do not know what he needs, so perforce you must think of his problems and difficulties and consider how best you may help.

The more you do this the less you think of him as a name or a cipher. He becomes instead a creature like yourself, with the same hopes and fears. Insensibly you get the habit of thinking of the other man, and this is convenient for him besides being good for your own soul.

Money is a very useful thing and I am very far from decrying its use, but it is not a solution to human problems, and all the gold in the world will not buy happiness, though it seems the fashion today to think it will.

When my nearest neighbour, aged 70 goes to milk my kye and gather potatoes a whole day in the October cold field, no money could pay for the service she does me.

The older I grow the more I feel with Robert Louis Stevenson that the "true services of life are inestimable in money and are never paid."

THE LOVE DARG

23 October 1945

Just now we and all our neighbours are busy with the peregrinating threshing mill. None of us has a farm large enough to warrant proper threshing tackle of our own – although many of us do possess small mills, which are mightily convenient in an emergency.

The trouble about these small mills is that, in the first place, they do not make a job of threshing barley, and, in the second, they presuppose larger and more rat-proof storing space than we have.

So, for these and other reasons, we remain for the time thirled to the travelling "millie", and so help to keep alive, though not very robustly, the last of the vanishing country love dargs.

In the old days the love darg was an integral and charming part of rural life. Clippings, dippings, hoeing, and hay all could count on the presence of helping neighbours. Hospitality was taken for granted, and in those spacious days a sheep was slaughtered to make a dinner for the workers while demijohns of whisky stood by to encourage them to ever wilder and more extravagant feats of skill.

These occasions were indeed the very stuff of country life. Old friends, new loves, and everyday enemies met there, and although the obvious reason of the love darg was to get the work done it was still infused with a spirit more gracious than the mere material end implied.

It is fine to be independent, but it is sweeter to recognise with humility that man needs the help of his fellows.

But alas, the love dargs are going, and the "millie" is almost the last of them. It, too, is losing its character. How can it be otherwise when instead of waiting upon the jade rumour, conveyed about the countryside by a travelling vanman, we take the telephone and make an appointment for the mill to come to us on a certain day.

The mill may or may not be in the vicinity, but the office in the town knows where it is, and can order its comings and goings. After making our appointment we ring the Food Office to ask for extra food.

Fancy a tin of bully beef instead of a side of home-grown mutton! And maybe we won't get even beef. While as for whisky – well, with a bit of luck we may get milk by docking the calves of their morning meal. Finally, we make a list of neighbours who will be likely to come to help and whom we shall help in return.

The list gets smaller every year, for death and age take toll among us and there are no young people to follow us in the hard hills. In despair we ring the much-tried and miraculously patient Labour Officer and ask for "furriners". He comes to the rescue, and our preparations are complete.

"ERE THE WINTER STORMS BEGIN"

26 October 1948

By making furious and desperate efforts last week we, along with our neighbours, managed to get the last of the tatties up. As a result we were all able to attend the Harvest Thanksgiving last Sunday in a somewhat more suitable frame of mind than if we'd had to be up and face the tattie digger in the rain and dark of the following early Monday morn.

We had hoped for a day or two's respite after the tatties, but the threshing mill is now on the go and Ed, complete with his twin darlings, the big International and its attendant mill, is on the rampage down the glen. That means we'll need to get going and see about things like tow for the buncher, bags for the corn, and extra rations for the extra hands. In spite of the feeling of being out of this and into waur, we are quite pleased to see the mill.

The backend is turning out blustery, cold, and wet and we're starting to feel that the beasts would be the better of new straw and a tickie bruised corn.

We've put off housing the stock as long as possible, but the milk cows are beginning to mourn now after the evening milking and gaze longingly at the roofs looking for in. I must confess that with squalls blowing up from an angry firth and the birch leaves swirling everywhere in furious tirievees, the fireside looks very inviting o'nights, so perhaps the cows better come home and put their feet up on the fender and take out their knitting.

All the same we'll make heroic efforts to keep all the young stock, except, of course, the very infants out all winter. We have grand shelter in the howe and plenty of it.

That according to the grieve is about all we have plenty of except he adds in his own sunshiny way, fresh air and water.

I must confess that I do wish we had better neeps. If we keep out the younger beasts that means we must neep them in the field and that in turn means fully more turnips per beast than if they were being fed in the byres.

It is true too that we have no lack of straw, but, oh dear, the barren months of March and April, these starvations of the hills, must be remembered and catered for. Do you think if I bawled at the poor offeecials in the D. of A's office in Elgin I might get coupons for extra feeding stuffs?

I doubt it, but I can aye try.

KNIGHT IN GREY PIN-STRIPE

28 October 1961

When I boarded the train at Glasgow he was sitting in the corner of the carriage, dashingly attired in miles of grey strips from socks to pullover, with a sparking white shirt and scarlet tie to point up the sophisticated colour scheme. Everything about him was suave and sleek except his eyes, eager and blue, gazing with wonder at his 18-year-old world. He overwhelmed me with kindness, arranging my luggage and doing clever and complicated things with the ventilation system in a split new British Railways carriage. The least I could do in return for all this old-world courtesy was to listen. So I did.

High Life

He was in the R.A.F. and was returning to Kinloss after a week's leave in his home town of Glasgow; and was he telling me? The R.A.F. was the life and especially in Kinloss. Whee! What a place. His voice was ecstatic. Now I have lived in Morayshire for a good slice of my life and Kinloss had never struck me as being all that gaudy. When I was a girl it wore the gentle air of ancient wisdom peculiar to the precincts of ruined abbeys set among fertile lands and sunny orchards. In spite of the huge modern air station it still retains its placid atmosphere. Or so I thought. My youthful cavalier saw its fields and woods quite otherwise. What to me where the haunts of the cultured and generous prelate Robert Reid, dead since the marriage of Mary Queen of Scots to the Dauphin, were to the boy the setting for pioneering adventure set in a thrilling key.

When he'd first come to Kinloss he realised his first duty was to get a gun to go a-hunting. True, he'd been born and brought up in Glasgow, but he recognised the urge of man's primordial nature which was to hunt. Being a child, too, of his age he had filled in all the necessary forms. Then he went off to the woods on the chase. He aimed – he shot – he killed – a rabbit.

Sweetheart

And as if that were not triumph enough he'd found himself in the very moment of high success, a sweetheart, a girl walking in the woods and crying. Because? Well, what do damsels cry for? So he led her chivalrously to safety – he carrying the gün – she carrying the rabbit, blood and all. I thought it pretty brave of her myself. I wondered too what the ghost of David Ist was thinking of the episode, for it was here centuries ago he got lost in the forest of the time and offered the Virgin a *quid pro quo* in the shape of an abbey if she'd get him out of the jam.

My hero was now fully launched on his exciting prospects. He had no intention of stopping at shooting and romancing; his next step must be to obtain a horse. "Like Jim Hardie" he sighed. And of course the whole thing clicked into position for me – the sleek pants, the gallant manners to a little old lady (me), the air of being at ease in a he-man's world in the country. There's more to looking at Westerns than the pundits think.

So I told him not quite how to get a horse but how to get in touch with pony trekking people who would help him to his dream. Did I think they'd let him always have the same horse? He'd like to get to know his steed; and did he need a licence and insurance. I could see him see himself galloping down the glades which the Forestry Commission call fire breaks, rescuing damsels who had annoyed their parents, shooting injins who were rabbits and generally acting with all the just aplomb of Wells Fargo's special investigator.

And all the time the wind would blow salt from the firth over the empty sands which I knew as stretching between Kinloss and Burghead but which to him were lonely Western deserts. I left him at my little country station, his eyes full of dreams and his enviable lot in front of him.

177

ALL KINDS OF FIREWOOD

We cut all our own logs with a circular saw rigged up with a belt which is driven off a pulley on the tractor. The sight of well filled wood sheds is a consolation for the lacerating mornings I spend when the saw is working. Even the blessed silence which succeeds the tortured shrieks of the saw is balm.

Most of what we burn here is fir or larch, although we try to save any straight bits of the latter to make into fenceposts.

Fir makes a splattery fire. It crackles away like the laughter of fools, but while it lasts is gay and warm. When the flames are gone, however, all that is left is a wavering white ash, and the grate soon gets cold again.

Larch is slower burning, and I fume when by mistake some of it is cut for morning kindling. But for an already burning fire which only requires replenishing, larch is excellent. It is inclined to spirt on to the floor, but when its first objections are over it burns slowly and methodically and leaves a heap of red embers which hold their heat for a long time.

Both fir and larch are better used dry. For a log to burn when it is wet one cannot do better for ilka day use than to put on birch. Indeed I think that birch is best used yet. Its clear blue flame singing its way round the silver coat of the log releases the sweet perfume which is the tree's gift to a raining spring evening.

Of course for something which burns magnificently when wet there is nothing to equal bog fir. We do not have a peat moss here, so do not come upon the great roots which are a by-product of cutting peat. But once we lived amid nothing but moors and then we found plenty of it.

It roots spread in every direction in the wet moss, and from the size of even its smallest tentacle we could see in imagination the greatness of the tree which it once fed. When we took it home, it burned with all the fierceness of the forgotten suns which had once nursed it into being. As it burned, a heavy spicy fragrance hung upon the air like incense and our great open fireplaces were filled with such leaping loveliness that it was easy to understand why our ancestors were fire worshippers.

They say that hard woods burn most satisfactorily and that old apple boughs make sweet and scented fires; for that I cannot say. I have, it is true, three gean trees which I am continually threatening with destruction. They grow all anyhow out of the ruins of an old stone dyke and do not improve the plot of land which we euphemistically call a garden. Indeed I blame them for the ills which finally blasted the rasps.

But my heart misgives me. For a few brief moments in the spring they gladden winter weary eyes with their blossom. A spray of cherry blossom against the green and evening sky of spring, a thrush singing *Gibbie Doak where hast tu been, where has tu been?* Well, I cannot demolish the trees. Then again in October their leaves turn to brighter fire than ever I could light – cold scarlet fire that warms the senses and the memory.

I know in my bones that long after I am gone the geans will flourish in the spring and flaunt the autumn out.

HOUSE WARMING

At last we've got the byre finished and the cows are for the first time enjoying their new quarters. If it had not been for harvest and tattie lifting intervening the work would have been done long ere this, but as it was we had to continue milking in the fold where we had to improvise binders and feeding boxes for the cows.

We were very proud of our faked-up arrangements. As a triumph of ingenuity over efficiency they were all that could be desired. The cows were certainly secured by the neck, but the rest of them could dance about quite a lot.

Bella, being well brought up, behaved best, although every now and then violent tremors attacked her, the result of Biddy's mesmeric and predatory eye. Bella is our main milk supply, and therefore she gets a little extra feeding which infuriates Biddy.

Even in the byre, confined to a single stall, Biddy teeters, but in the fold she waltzes. Frank milks her to an accompaniment of violent Italian which rouses Charlotte to nervous crisis after crisis. This in turn agitates the placid Rhoda who pretends that the exaggerated shadows of the cats dancing pavanes around the lantern are dragons' ghosts.

Altogether milking time in the fold was not without its tension.

But to-night how differently matters stand! We were not sure at first whether the cows would take to their new ultra-hygienic quarters. We wondered if they would be overcome with nostalgia for causeys deeply imbrued with strong-smelling ammonia, and what would they say to no haiks full of straw on the walls, but simple cement feeding-boxes?

I am glad to say our cows move with the times, and when we went to milk they received us joyfully. It was laughable to see their innocent pride. They looked so exactly like hard-working women who had spent years running shockingly inconvenient houses, with attics and basements and no kitchen sinks, and were now translated to a modern labour-saving home.

Instead of getting all set to argue that they would prefer to be milked at the other side they actually made way for us. A sleepy air of contentment hung over all.

Bella ate her mixture in peace, hidden from avaricious eyes by the walls of her new-scrubbed stall. Biddy, standing alone in her single, specially long stance, came the socially superior stuff. Thank goodness the days of mingling with the polloi were over. Let the mob dwell together in double stalls, but she was accustomed to different ways. Rhoda is sunk in complacent thoughts of approaching motherhood, and Charlotte grudgingly admits that there are now no draughts, and there actually seem to be enough places for humans to put things down without upsetting them.

We, too, are grateful for a smooth passage where we can run up a barrow of feeding stuff without having to labour back and forth with scows, and the sight of a properly constructed graip which can be easily mucked gives us more solid satisfaction than many a lovelier view. So, for once, we are all happy.

NOVEMBER NIGHT LIFE

3 November 1953

Now that November is upon us, night life in the glen hurries on its glittering way. Reels, rurals, badminton, and the Women's Guild succeed each other every night of the week except Friday, when everyone whists and dances in aid of funds for the aforementioned activities.

All the same, all is not fun and games for everybody. Ask any haggard secretary on any of these bodies what they think. Mind you, I don't think the badminton or the country dancing officials have any real grouse, not in comparison with their corresponding numbers on the rural or the junior farmers. After all, if you dance reels you just dance reels, if you play badminton you just play the game.

You don't have to provide a split new programme for every meeting, which is what the other unfortunate secretaries have to do.

Away back in the leafy month of June you see haggard rural secretaries penning away for dear life trying to compile a syllabus for the coming winter. "So difficult to get speakers, so dreadful to get demonstrators," they moan. And so they write letter after letter to folk who they think will be sympathetic.

Let us now suppose that the cry reaches a kindly ear. The ear consults its diary, for no amount of sympathy will allow the same body to be in two different places the same night. If there is no prior arrangement then the speaker takes a further look at the date. What like will the weather be in the depths of Glen Macwhither in the second week of February.

However kind the heart it recoils in horror at the thought of having to be rescued by St Bernards whilst on the way to demonstrate how to make felt slippers out of old hats to the women's club. So a correspondence between speaker and secretary gets up and alternative dates shuttle back and forth. So nerve-racking.

No wonder then that the job of secretary is shunned. All the same a few simple obvious rules would help a lot. The first rule is for the secretary always to enclose a stamped addressed envelope. Always tell the speaker that arrangements will be made for the journey back and forth. If overnight hospitality is necessary make it sound as if the speaker will be an honoured guest. If you sound as if you were conscious of having a favour conferred rather than t'other way, you'll be surprised at the favourable responses.

WINTER

WINTER WITHOUT DISCONTENT

4 November 1961

With Hallowe'en past and summer time officially dead for a week, we cannot delude ourselves much longer that the winter session is not imminent. True, the birches still wear their vestigial gold and the embers glowing along the gean tree boughs are not extinguished; but come one more gale and all will be quenched. There will not remain one forgotten rose in the little garden; only the tashed remains of dwarf Michaelmas daisies. Our magpie will shelter in the pinewood at night and the pheasants flaunt and skreigh above the rusty filings made by a fallen larch tree needles. A fox will bark somewhere in the middle of the dark moor and a roe will grunt from the planting, while little owls prophesy storm under the reeling moon.

However, as the inexorable routine of upland winter farming takes over the measure of our short days, we shall not find much time to regret the passing of summer. Already we are busy with the problems of keep. The sight of two large pits of silage is comforting but no one can tell how late spring can be in such regions and we are better not to gamble. So we fidget around arranging for supplementary rations.

Draff and stock feed potatoes are our stand-bys – or they were until other farmers also got the same idea. Now if you want draff which is a by-product of whisky making you must put down your name for it away back in the summer time and then take your place in the queue. The other day we saw a lorry from the plain of Buchan skelping through Forres with a load of draff which the driver had collected from a still in Ross, and one could not help wondering how expensive such feeding must ultimately turn out to be since the freightage nowadays costs the earth.

On the whole we may be better to opt for potatoes. This will certainly please old Morag who is bringing up two calves and never lets us forget about her claims for special treatment. Yesterday in the middle of a fierce though flying shower she pursued Dod into the steading and bawled at him till he was forced to produce some hay which she alleged was very dry and was there not a tattie to go with it? The big softie went to the household bag and stole her a boiling. It is not for nothing that I hope the stock feed tattie situation will be fairly easy this year.

But we do not have only cows to worry about in winter. We have wintering sheep as well. They came in at the end of September and had a good dry spell to get accustomed to their new quarters and our hill. All of them behaved beautifully except seven who have turned out the most stravaiging brutes in the Highlands. Every time the phone goes we fear that some one else is ringing up to tell us that the seven have turned up in some one else's pastures. We have been round all our fences and cannot discover where they make their escape.

My own theory is that they dematerialise themselves on one side and come out as ectoplasm on the other. Dod says that they have just got a notion to get a hurl in the back of our estate car because when he goes to take them home he takes them by car. He has a little box with sawdust in the bottom and when the wandering sheep see it they line up and jump in. Once aboard they tap on his back to tell him to get going and breathe heavily as the scenery flashes past. He has tried them in every field in the place but in vain. They say they don 't like our grass. They prefer moss and birch tree bark. Perhaps they think they are reindeer.

VILLAGE HALL ECONOMICS

5 November 1960

We are in the middle of one of these financial lurches by which we keep our village hall in its usual state of precarious equilibrium. The county council, whom we have approached, have sent us a form on which we are to write down the details which will entitle us to get help in paying our current way, but unfortunately the form is very difficult to fill in.

The wretched thing is out of date in the first place, but the council is so thrifty minded that it cannot bring itself to burn it and ask for a new one; which means that we must do as best we can and rely on some one writing a cogent covering letter. The committee regard the proceeding with deep suspicion, and I know that if I take my eye off them for a minute the glen will find itself involved in yet another sale of work.

Slow Count

The thought of having to bake a cake and look out more jumble lends spurs to the sides of my form-filling-letter-writing intent so that I demand in no subservient tones that the secretary provide me at once with the number and kind of the societies which use the hall. Giving a pitiful look at her co-members she slowly counts out the activities that go on in these precincts. As she names them, all present, severally and together, say a few pithy words about each society of which the speaker is not a member. Life in a depopulating glen, inhabited albeit vestigially, by two inter-related clans is not without dramatic interest.

By the time we have agreed to differ in our bitter way about who should or should not play the accompaniments at the next concert, we

189

have also had occasion to remark on the hellish racket the hall piano makes even under the most mellifluous fingers. So I write down that this hall is in use every night of the week except Sunday, but as the infant Sunday school meets there on Sabbath mornings, Sunday, too, has its quota of use.

Contempt

The next question deals with the age of the members of the various societies. A deep silence, indigo with unexpressed fury, hangs on the horrified air as each person envisages going to the Rural and asking to view their birth certificates. In the end we treat the query with the contempt it deserves and draw a good emphatic stroke in the answer column.

Foolishly I then propose to fill in the ages of the two badminton clubs with the general term "teenager." A huffed voice from the back tells the general listener that many adults also play badminton and they have no wish to be associated with the youthful riff-raff of the glen.

Like Eliza on the ice, I leap from one hazard to another. How much was the instructor paid who taught last year's country dancing? Did the local lass who played for the class get a fee, and if so, how much? Two factions are at once present, each with their own candidate for the piano fee as well as for any kudos so attaching. Tempers show ragged at the edge and nippy things are said about home lessons, the imminence of exams, and the difficulty of transport. How indignant we could become is governed by the increasing chilliness of our feet and we glance sourly at the paraffin heater, throwing its economical warmth at a circle 18in in diameter.

Electricity

And that brings us all together again over the cost of electricity. We may disagree about the prompter in the dramatic, but, by heavens, we unanimously think the Hydro Board are Shylocks. It serves them right that we have just got around to installing a gadget which will cut the electricity in the hall by half. It is a pity that the confounded thing cost so much, but in three – or is it five? – years we shall be able to spit in the eye of every meter reader in Morayshire.

On which uplifting thought the meeting ends.

THE CARAVANS GO BY

6 November 1945

The caravans went down the road to-day, lumbering their laborious way from the high hills to the beckoning sea. The recent storms had gone and now the world shone again in brave autumn finery.

There was just that touch of frost in the air which makes all colours doubly bright and endows the landscape with a radiance half like that of jewels, half like that of flowers. Sky, painted woods and silken sea were lovely sisters in a shining world.

There was quite a trail of caravans making their way down to their winter quarters in the plain and they made an impressive spectacle as they wound over the bridge. Gone are the days when they wore liveries of scarlet and green and relied on a horse to pull them. The ones we saw to-day were painted a douce shade of reddish brown and powerful motors sheltering under the same roof as the living quarters chugged easily along.

And yet, for all their sober appearance they had not lost the power to enchant this observer anyway. In a world distracted by rules and restrictions, the sight of a caravan seems to say that somewhere men are unfettered. Ordinary men must stay in one place and earn their living there, but there are folk who can range the world in search of wealth and adventure.

I remember one halcyon summer before the war meeting a family of caravan dwellers near Melrose. In their case van and motor were divided and they managed to combine the best of both worlds in the oddest fashion.

During the day they toured the country in an enormous, flashy

Yankee car collecting "broke" from the sheep farms. (Broke is the odd and dirty pieces of the sheep's fleece which spoils the rest of the clip when it is included). At night they return to a supper cooked by the mistress, kerchief on head, on a tripod over a fire burning palely against the summer sky.

A little girl ran piping amongst her elders, she had an ivory coloured face and great black eyes. Her ears were pierced and she wore earrings of coral, turquoise and pearl.

A casual friendship grew up between the mother and me, chiefly over the unromantic subject of cod liver oil for children. The little girl was not strong and I suggested this remedy.

Then one night I was invited in to the van itself. Hitherto our encounters had been alfresco.

It was exactly as if I had strayed into a greatly magnified edition of my grandmother's rosewood workbox. Where everything did not shine with the satin of polished wood, it shone with the satin of real satin. A silver lamp hung from a wall and the bed was made up with a satin coverlet. Two silken pillows embroidered fantastically guarded the dreams of a gipsy baby.

I came out into the air again feeling I had been in Araby.

THE WITHERING YEAR

10 November 1953

We light the hens' lamps a quarter of an hour earlier every night now for at this time of year the daylight slips away quicker than water and we no longer have any compensating sunset glow. "At one stride comes the dark."

All the same the backend has been wonderfully mellow with the painted trees standing still against a timeless sky. Sometimes in such gentle days when the farm work goes on without hindrance you wonder if a good backend isn't worth more than a sunny summer. Then you remember the consequences of the rain and gale of last summer.

We are still getting loads of rakings off fields where harvest proper ended ages ago. What a nuisance these rakings are too for we don't quite know how to get the best out of them.

As they are very full of corn we thought we'd shove them into the deep litter where we trusted the hens would work amongst them. We thought wrong. They merely choked the shed and the hens acted as if they were stranded on an inhospitable mountain top. So we had to take the darned things out again.

Then we let the hens out to scratch amongst them in the yard. Oh, they scratched alright and either went off the lay or hid their eggs in impossible hidey holes. Eventually we decided to build our embarrassment into stacks and leave them for the outwintering cattle in time of storm. I hope Dod will build the stacks where we can't see them to be reminded of the loss.

Yet with all the good weather we are constantly admonished of the imminence of winter. A nearby neighbour has just limed one of his fields, and the stuff hangs like snow on the ground and rests on the branches of the trees just as snow flakes would do. We feel no surprise, therefore, when we lift our eyes to the farther horizon and see the mountains beyond the western sea blanketed in fresh white.

The firth itself is a cold steel blue and the air is sharp with the lances of frost. One feels one's years like icicles in the blood.

The wintering sheep come down out of the hill at morning to the fields near the house. There are not so many of them as there used to be for the forestry folk have taken the rough grazings to plant. The farmers mutter away but accept the inevitable.

NOW FOR NOVEMBER

11 November 1947

Now the time has changed and mornings are full of pale, chill light so that we do not need lanterns in the byre, where the cows stand benevolently bewildered by the sudden change of time-table.

Now night comes quickly and we must cease work early to pick our dusky way over the muddy cobbles homewards to the windows where the evening lamps are lit and beckoning. Now byre-time at night is long-drawn-out once more, while the lantern bobs between byre and barn and dairy.

Now the slow moon, mist-entangled, mounts to the heavens to wander her mild way along the paths of night till morning finds her pale, dishevelled, out of place in the sky already menaced by winter.

Now the cattle come in at nights, although they still go out every daylit hour to eat the turnips already spread for them in the garnered fields. Now the starlings chatter in the steading roofs, but their clamour is soon drowned out when the potato sorter starts its interminable clatter.

Now the riddles shake and the elevators churn till the potatoes pass through and over and find at last their way into the appropriate sacks.

Now farmers try to remember to notify the Department of Agriculture of their intention to start this work, while not forgetting to announce to their own seed association that they have so many tons of seed ready for inspection.

Now stationmasters tear their hair out as they try to get wagons for the finished seed.

Now faithful collies herd flocks of sheep new come from market to their new homes, where great bright sows of straw announce that the mill has come and gone.

Now October gales are done and beech leaves lie in russet drifts by the river and the bridge. Now the larches spend their needles, though the oaks, in contrast to the ash and elm, are still a sturdy Sherwood green.

Now dews lie heavily on the feathery dead grasses and hang on the frail silk of spiders' webs spanning the trails of briar and bramble.

Now the last red haw shines out, singing like a bugle that though the year is done time is not dead but sleeping, and when she wakes her eyes will open on the spring.

THE PROPER HAT FOR THE JOB

11 November 1961

We had the first powdering of snow this last week, which, yet left such a nip in the wind that we fled to the cupboard for extra woollens to help us withstand the blast. Headsquares, practically warm from the sheep and lined boots allied to our usual tweeds and twin-sets will see us womenfolk comfortably, if monotonously, through till Easter; and if we are inclined to bemoan our own douce unimaginative attire we can always comfort ourselves with looking at the dashingly unconventional garb of our husbands.

The more it snows and blows the more picturesque the countryman becomes. How buccaneering he can look when he dons the old Army coat he and the tractor share with happy impartiality. How ingenious he is when he carves leggings out of the remains of ancient Wellington boots. Looking at a prehistoric Home Guard tunic tastefully lashed with binder tow, one cannot but wonder what the new Ever Ready uniform will eventually look like after it has encountered a resourceful farmer negotiating snowdrifts and floods.

But of all the accessories that go to make the well-dressed countryman none is so impressive as his hat, or should one say his hats? Every man about the moor has an extensive wardrobe of hats and each is worn on appropriate occasions. When he is young, of course, he has not yet collected many. He, like his town counterpart, favours rather the fancy hair-do topped with a skid-lid.

But even the youthful rustic must have at least one orthodox country hat. This is the one he wears while driving the tractor. It is really an engine driver's cap with a shiny plastic brim and if it did not belong to the young countryman would be sober enough.

Thus wind from the east means to tilt to the left and a tempest out of the north means the headgear flat on with the snout to the front; while a blow-up from Cairngorm to the South forces the wearer to turn his cap back to front and allow the wet to drain off the plastic down between his oilskin-covered shoulders.

As the youth grows older he discards this frivolous affair and goes all out for a Harris tweed thing with a full crown which enables him to use it for drying a plug or wiping superfluous udder salve off a cow. His wife takes a poor view of this and threatens to throw it out and then the farmer takes to a deerstalker. The latter he uses entirely for wild forays on the hill among the sheep. Many waters and storms have softened the brims both fore and aft but the ribbons tying the earflaps on top remain curiously jaunty.

This is in vivid contrast to the strings that secure the neck flap on the ski-ing cap he stole from his wife. Made of hard wearing Grenfell cloth it is used mainly for haymaking in a Scottish summer and is apt to get sweat stained, in which case it is cleaned by the farmer scraping it with his pocket knife and getting up the blindside of one of his family to wash it surreptitiously when his wife is not looking.

Extremely romantic is the helmet made of waterproof and lined with woolly khaki. This is worn when driving the tractor on the occasions when it is not convenient to have the tractor hood in place. The helmet is constructed on the ample side so that it presents a flowing line round the back of the neck reminiscent of a Roman legionary.

There is no room to tell of the function of the beret, the Balmoral, or the Balaclava in the countryman's life, but all play their happy and significant role.

THE EXERCISE OF HOSPITALITY

12 November 1946

It is an automatic instinct in the hearts of country dwelling folk to provide hospitality for all who come to their doors, and nothing so shames the housewife in these parts as not to be able to give food to the stranger at her gates. As a community we are poor but not so poor that we cannot share. And to tell the truth there seems to be a blessing in what we have for we can always give a meal and never miss it to the man who comes from anywhere.

It is a pleasant and gracious privilege and one we cherish. Yet in these difficult days it is a pleasure we have to pay for in ingenuity and skill and when the mill comes round we find our contriving and scheming get full scope.

Just now the countryside reverberates to the excitement of the itinerant threshing mill, for the good weather has fired every farmer with the hope of getting this important day past as long as skies are kind and roads clear.

The result is that every farm kitchen must be prepared at the shortest notice to provide three good meals for anything up to a couple of dozen folk.

It is true that a beneficent Food Office realises our needs and provides a moiety of coupons against such emergencies, but the allowance is very small, and it needs the greatest care to make the most of it.

My heart bleeds for the poor shopkeepers in Forres, for we hill folk descend on them with the light of battle in our eye and demand the uttermost crumb for our permits. They do a noble best, as you may judge from the fact that my baker opened his shop at 10 o'clock at night to fulfil my own mill order.

Sometimes, of course, we do think the exercise of such hospitality is rather expensive in planning and general trauchle, but when we see the long tables full with our guest helpers we know that it has been worth every bit of our endeavour.

A housewife asks no more than to see the uttermost drop of her broth supped down and the very last pancake gone.

As the swift November dusk wells up out of the valley the last helper disappears behind the tractor-driven mill, and the housewife watches them go, calling her thanks and farewells after them.

She is almighty tired, but, mercy me, she is satisfied.

HARVEST (NEARLY) HOME

12 November 1960

After one of the dreichest Octobers of the century November has decided to make amends, in token of which she set to one night this week and swept every cloud and vapour from the sky, after which she whistled up a tingle of frost fit to burnish every last star in the firmament to blazing brilliance and cut a few million light years off Orion's age. Nor was the sun slow to follow such a noble example for he rose purposefully at dawn and sent the morning brave and blue out over the firth to the hills of Ross where Wyvis stands like a Celtic Soracte wearing a candid crown of shining snow.

Change of Fortune

The poor half-drowned humans can hardly believe their change of fortune and draw incredulous Wellingtons from the mud and slime of weeks of flood before they contemplate salvage operations. But our indecision cannot last long if next Sunday is to be the harvest home festival. With an effort we think we may just make it and we hope that the Almighty will understand if our hymns do not sound quite as heartfelt as usual and are more suggestive of relief than joy.

I see some farmers are quicker off the mark than others, for already two huge scarlet combines have gone lumbering up the hill to Strathspey, where the oats lie withered and twisted in the ugly drookit fields. I wonder if our neighbour has managed to get his combine out of the bog where it has been slowly drowning since the last flood but two. If he has, then there will be three machines on the go and surely everyone will at least secure their grain, though the straw may remain to moulder in the park.

Valuable

The trouble, of course, in this countryside is that straw is almost as valuable as the grain, since it is used for bedding beasts as well as for feeding them. Since we ourselves stopped harvesting in the conventional way we have arranged other feed than straw for winter keep and have made out with sawdust for bedding. Thus when the rain came we were able to fill in the time pluistering in hill drains and running down to the timber mills in Forres for bogie loads of sawdust.

Naturally our farming friends took this as yet another instance of the peculiar way in which we farm; but we begin to notice that sawdust is becoming scarcer than of yore, and if we want to make sure of it we must phone down and reserve our share. Could it be that other farmers are following our example? It would be a little odd if one of the consequences of this difficult year were to be that sawdust, which has been so long an encumbrance to the timberyard, should become a useful commodity. There is a feeling that sawdust used on the land is bad for it, but we have used it for a top dressing for grass with excellent results.

Cheerfully

Humans are not the only creatures to rejoice when the sun shines. The beasts of the field, feeling the warmth and seeing their paths clear once again, add their meed of praise. The wintering sheep stop looking like drowned rats and the autumn-born calves dance in excitement on the dry hillocks. Birds, too, cease their whickering in the undergrowth and fly cheerfully about the feeding troughs in the barnyard. In the plantations cock pheasants soar in splendid Eastern plumage and challenge the native grouse to match the iridescence of their tails.

Far up in the sky the geese jangle past. Long after they have disappeared over the hill to the south we hear the wild free sound of their wandering. For weeks we have heard them but today is the first time we have seen them. Above the pale gold of the willow tree by the burn, above the drying embers of a wasting autumn the geese pursue their untrammelled way. I hope we'll all have drier feet the next time they come back.

GOOD-BYE TO AUTUMN

13 November 1945

Autumn has said her last long golden farewell, and winter's vanguards crowd the horizons. Storms of hail and rain sweep the hills, and angry blinks of sunshine interrupt them.

Great clouds of birds are blown down the sky before the wind. One moment they are visible, and then the lowering trails of mist hide them.

Today the rooks went by the window.

> *On the first of March*
> *The crows begin to search,*
> *By the first of April*
> *They are sitting still;*
> *By the first of May*
> *They are a'flown away,*
> *Croupin' greedy back again*
> *Wi' October wind and rain*

When the wind goes down again they'll settle in the larches and call derisively to the domesticated hens, who heed them not. But our duck, when the gaggles of wild duck go by, stops to lift her head to see them go. Some almost stifled impulse in her worldly heart bids her listen, but, alas, her heavy wings and body keep her chained to the fleshpots.

Not for her the lunar rainbows of the starry North; not for her the bewitched beauty of Arctic wastes. She beats her short wings as if she were wringing her hands, and quacks distractedly after me, furious that the hens reach me first.

The sky with its siren call to freedom beckons, but the smell of hot mash is more insistent. Keep your adventure. Give me a full stomach, a warm bed, and a muddy pool; only sometimes when the sky is tangled with long skeins of duck flying seawards is the price too high.

Two yellowhammers have taken up their winter quarters beneath the scullery window.

> *Half a puddock, half a taed,*
> *Half a yellow yorling,*
> *Drinks a drap o'deil's blood*
> *Every May morning*

Their summer's song with its half-dozen short reedy notes hurried out and then followed by a long thin note, is now mute. Our ancestors said that the words of the song were "Deil, Deil, Deil tak' ye" and according to the macabre legend the drop of devil's blood which the bird drank every May morning was used to paint its eggs.

Certainly, to look at these graceful golden creatures with the sunshine lingering in their wings one does not have the feeling that they belong to the nether regions. Soon we shall hear the wren and robin, and then, indeed, we shall know that winter is here. In the meantime, farewell Autumn. Winter be kind, be short.

PREPARATION FOR THRESHING

17 November 1953

We are having the threshing mill tomorrow, a prophecy with the usual reservation as to weather and Ed's chock-full programme. If he has been delayed anywhere else this week we need not look for him till the middle of next week.

Since the relaxing of food restrictions the threshing mill is not the nightmare it was once, and if the farmer's wife enjoys ordering sugar and meat to her real requirements, the Forres tradesmen, too, must enjoy it. The way they rallied round and made your mill worries theirs remains a green and grateful memory to us all.

Nevertheless, the whole question of feeding mill-hands got more and more difficult, and the only way to get round it was to take the mill for a half-day. Actually this worked out well, not only for the commissariat but for the farmer and his scanty staff, since it meant that they got their own essential jobs done as well as lending a hand to a neighbour.

For most of our neighbours this arrangement still persists, but, alas, for us, we are well within the snow-belt which the rest of the glen escapes, and this means that we must always take into account the state of the roads when we need a mill in winter.

As long as the going is good we must take a full day, and hope our understanding friends will forgive us.

Hands in these parts continue to be difficult to find, and we are glad to hire a baler to deal with the straw. Baled straw does not need half the folk to cope with it that bunched straw does.

It has the further advantages of being capable of being packed into much less space, and can be pretty well contained in the barn, whereas the bunched stuff depends on skilled building to make it into a sow which will stand the stress of wind and weather.

Believe me, it's a whole lot nicer taking a bale of straw out of the barn on a windy day than tripping off to the field for a load of the stuff what time the Dava tempest rages.

We thought it might save phonings if we hired the baler along with the rest of the Department's tackle, but Bert, who arranges these things, near grat when we mentioned it.

"We have one baler and it's far too hard worked" he said, so with remorseful hearts thinking sorrowfully of the baler's iron jaws being worn to the gums we hastily said we'd get another baler not so sorely tried.

CRYSTAL BALL METHOD OF FARMING

18 November 1961

We sold the first batch of our spring calves last week and have been absorbed ever since in the inevitable post-mortem which accompanies all our farming enterprises. Should we have sold our beasts when we did; or earlier when the first big calf sales took place with all their ballyhoo, horde of buyers from the South, and general air of plucking a premium bond from Ernie's vitals?

Were we after all better to have waited although the pattern we long to discern never emerged? In an industry hazed with imponderables like the "Charley" bulls and the looming Common Market perhaps no pattern was possible.

I'm all for the crystal ball method in deciding future action in the farming sphere but Dod revolts at such irresponsible shelving of logic and insists on working out the difference in costs between the September calf sold in the following June and the calf dropped in February and put on the market in October.

The September calf will be bigger and older and more beautiful. On paper it will make a higher trade; but will the money it leaves really be so much more, or is it a chimera? The February calf, unlike its elder brother, has a mother who will need no boosting to keep up her milk supply. The obliging creature will fend for herself, and the season will provide fresh growing grass which, as all the pundits tell us, is not only the best food for cows but also the cheapest.

Well why then do we bother with the September calf? My reeling mind can only reflect that selling such calves, resplendent, black and satiny in June, is nothing but a status symbol of the pastoral farmer – perhaps the modern version of the ancient Golden Calf?

So the process of the post-mortem goes on its involved way. As well as known facts it must also consider the eccentricities of a market which can add or lop off as much as eight pounds to a beast for no known reason. Don't ask me why heifer calves for the past two years were sold for little more than sweetie money and then this year their price suddenly soared.

While this suits us we'd be glad to know what moves the buyer. Also what dusty answer gets the soul, when hot for certainties in this our life. The auction ring remains for ever unprëdictable.

But November routine on an upland farm soon imposes itself on all speculation and accountancy, and the short grey days have their own compulsion. So have the cows on whom we ultimately depend for our income. They are quick to point out their likes and our duties. Have the slurry aprons leading into the silage pits been scraped they would like to know? Self-respecting cows dislike getting their feet "dubby" just as much as humans do, and that is one of the reasons they now enter the cattle court, which has been levelled and cindered, with much more pleasure.

And there is the matter of mixed menu. Do we really mean to cut out the cows with the youngest calves from the lush rape mixture in the field next the hill? There is no use us going all scientific and insisting that newly calved cows should begin the day with a breakfast of hay when everyone else is tucking in to green green rape.

Upset their calves with too much milk did we say? Apparently ignorant humans have never heard that a little of what you fancy does you good and to pot with balanced diet and roughage – the calves will survive a little tummy upset surely. So saying the greedy brutes make a breenge past the gate and leg it up the pass to the rape field. They know that Bob, the collie, recognises the impossibility of arguing with cows plus calves. We know it too but hope the electric fence will teach them their place.

STANDSTILL ORDER FOR THE FARMERS

19 November 1960

It only needed the outbreak of foot-and-mouth disease at this back-end to underline the bad eminence of 1960 in disastrous farming history. We could have tholed the great swatches of a ruined cereal harvest still lying wasting in the mid-November fields; we could have gritted our teeth and kept going through acres of waterlogged potatoes which will never now be lifted; but it is going to require every last ounce of resolution as well as every last penny to let us survive this final misfortune.

Obviously a standstill order extending from the Caledonian Canal to the whole of the South of Scotland must bring a weight of cold sickness to every farmer's heart; but to us as livestock producers the order brings many more mundane worries. We are not recipients of the monthly milk cheque, nor do many of us any longer depend on the weekly egg slip. Our farming accounts are geared to the annual cashing of assets which are due to be sold in the last three months of the year.

No Qualms
Our merchants accept this economic fact and do not bother us unduly for payment, financing us without a qualm through the rest of the year till I begin to think that some one should write a grateful thesis on the place they occupy as bankers in agriculture. But even the most trusting of firms want to get payment sometime, but if the farmers cannot get to market then there is no money and debts remain as debts. Nor can we follow the example of the famous Turra Coo and present our creditors with a calf or two in lieu of cash, for the standstill order means exactly what it says.

That farmers and indeed all country people accept the order with such absolute co-operation is a measure of the anxiety which desolates us and makes us cancel every kind of contact, social and business, which could lead, however indirectly, to the spread of the scourge. Many farm jobs which are dependent on outside contractors will now have to wait a more auspicious time, so that we can foresee a wild rush of work when the standstill order is lifted. For ourselves we are not unduly put out that the limespreading will have to be postponed, but we'd have been glad if the renovations to the steading gable had been further on.

Fellow Feeling

Stuck as we are on every side we have more than a fellow feeling for the boys who make their living transporting beasts to market, and we do not forget the tradesmen in our country towns who must suffer their loss too.

But little can be gained by moaning, though no one can expect the man who has lost all his beasts to be capable of any feeling but an empty despair. Those of us who are fortunate enough to escape should surely use our sad compulsion to working out some kind of control to prevent a similar occurrence. Perhaps because we live in a scientific age we think that all ills on farm and town can be cured with scientific formulae, and though we should laugh if you were to tell us that our idea of the modern cattleman is a bloke with his waistcoat pocket full of hypodermic syringes each filled with antibiotics, you would not be far wrong.

Truths

But while we wait for the brains at Pirbright to discover the magic prescription which will conquer foot and mouth we can ourselves do something towards helping. A good hard look at our marketing might provide some useful if unpalatable truths. Do we ask the calf dealer where the black calf for which we are paying a lot of money came from? Do we, anyway, think it good for beast or farmer or to the industry that beasts should be hurled from market to market in order to catch better prices? And deadlier diseases?

THE GOING OF PANSY

23 November 1948

Pansy is our black and white cat and her behaviour is the occasion of this sad story. Pansy came to us, along with her two children Patch and Donna, from a farm in an adjoining strath, and at first we could not have wished a more virtuous mother. Her days were blameless rounds from the morning saucer of bland milk to the evening bowl of gently steaming porridge.

How admirably, too, did she divide her days. There was that first toilet when she shrugged herself into her newly pressed black housecoat and fixed her freshly laundered tucker. This was followed by the care for her children, their toilet, their breakfast, their instruction.

Mousecatching came next, followed by rat menacing in the barn, and if by chance you met her of an afternoon enjoying an elegant dish of tea with our orange tabby you felt glad that her busy day yet found time for the graces of leisure.

You can see, can't you, that Pansy was not the ordinary type of cat, so diligent, so discreet, so considerably gooder than truth.

Alas, that approaching age should make me so cynical, but the fact remains that when the grieve found Pansy blinking yellowly in the brief November sun on a rabbit-ridden hillock on the moor, I was not surprised.

Oh she carried it off well and asked the grieve to recover her tatting for her. She alleged it had slipped down a rabbit hole. But the seed of suspicion which had always been there began now to germinate.

There was the question of Pansy's fourth leg. Did I say she had only three legs? Perhaps I didn't like to mention it. Well, where had Pansy lost her limb?

Could it have happened when she had been having an unladylike prowl round the gamekeeper's traps? Horrid thought! Still, how had she lost it?

Then one morning she did not come for her milk, and three hours later I found her abandoned in dissipated slumber in the chaff of the fold floor. Opening one slit eye, she leered at me before she could recollect herself. It was useless now to try and make excuses.

My illusions were quite gone, and by night I found that Pansy had gone too. But not entirely for good. She occasionally flags His Majesty's mail and pays us a call.

She lives now on a neighbour's farm, and when she wants to annoy me she comes round with the postman. Sitting beside him, she gives me a sideways glance. "Girls together?" she insinuates.

I don't like it. I'm respectable, I am.

"GOVERNMENT GRANTS IN AID OF ......"

26 November 1960

It is a pouring wet November morn, and Dod and his collie have just taken off into the grey deluge to inspect the completed hill drains and cogitate on new ones, yet unborn, un-gushing. If it had not been for the standstill order because of foot-and-mouth precautions I expect the department's expert on hill drains would have been one of the party, but he is chained to his desk somewhere in the romantic confines of Inverness. Not that the poor lamb needs to be here in physical fact, since by this time he must know by heart every ditch and shoggle bog of our 800 acres of hill, while the stench from our sulphur well must ever be lively in his olfactory memory.

Dod when last beheld was carrying a large broad plank under his arm, while he wore a hammer and pliers in his belt; he evidently expects shipwreck in the wild wet moor, while Bob, his faithful hound, says he is practising life-saving on the antrin mountain hares.

Crusade

In spite of appearances we are not all that thirled to hill drains. It is true that in the past few years we have dug miles and miles of them and can say our "twenty-two yards one chain" linear table backwards, since a chain is the unit of drain measure. What has moved us to these messy Herculean efforts is Dod's private crusade for a higher and holier hill farm. Therefore he feels it a sacred duty to take advantage of every stricken Government grant for farm improvement – an attitude I find illogical since he also regards the acceptance of subsidies as immoral. Subsidies are what the Government pay you after you have been seduced into vast expense in improving your farm.

There are 13 grants-in-aid and we are to work through the lot – if we live long enough – and if we can go on scraping sufficient pence together to launch out on yet another scheme. What we do when we come to the bracken eradication one I cannot think, since we have very little bracken, but probably Dod will import some just for the sake of his Aberdonian principles.

Magic Word

The grant for hill drains is one half the "approved" cost, and "approved" is the operative word. But not all grants pay the same proportion of your bill. As well as the drains we are running another plan under the Farm Improvements grant and here we pay two-thirds while the Government pay the last third, and here again the magic word is "approved." Under this second scheme we are hauling down gables in the steading so that we can work tractors and dungloaders with more ease than at present. The blokes who built the cattle court here could envisage nothing more modern than a spade and a caschrom for hand-ploughing. We are also putting drinking bowls into a byre, and if that is not all mod cons for hill cows it very nearly is.

So far we have got the bowls in and I hope the rest of the plan will work out more smoothly than this bit is doing. The plumber assured me that beasts knew by instinct how to work the drinking bowls; he had just done a similar installation in Dallas and the stirks there walked cheerfully into a byre and at once pressed the button. I thought to myself that such sophisticated stirks probably thought they were phoning and were looking for a rake-off from Button B. Anyway our cows are not nearly so clever and you have to take Maudie by the hand to the bowl while old Morag from Skye says she prefers the water in the burn which comes straight from heaven. A pity there is not a grant in aid of educating cows.

WHAT'S IN A NAME?

30 November 1948

We sold a bunch of back calvers the other day and as a result I had to make up the Movement of Stock Record. The Law compels you to keep this book and it is as well to keep it up to date.

As I noted that Rosa had gone here and Angeline there, while Sara and Rachel had found still other homes, I reflected, and not for the first time, on the usefulness of names for the beasts. It saves such a lot of vagueness and muddle if you can say "Teresa" instead of 5975 C 14 or, worse still, the wee heiferie with the black face and the white hind leg.

Some years ago we purchased three heifer calves and it so happened that at the time I had been re-reading the Brontes. Irresponsibly I named the calves Charlotte, Emily and Anne. Thank goodness, in the light of what happened, there wasn't a stot for me to call Branwell.

First Anne went into a decline and died with pious resignation. "Ach, poor brute," says I. "A bad doer," and thought no more.

Then Emily, who was a black beast just the kind you like if you are trying to put on the beef, began to yearn after the moors and wuthering heights in which this farm abounds. Instead of staying in the lush home pastures she took to going for solitary walks on the hill.

I took an ill will to the black-browed introvert and sold her. I believe her new owner called her Maggie and she throve into prime beef in no time.

Charlotte I still have. She is red haired and passionate and obviously has a grudge against genteel service. I never like to ask her to demean herself by letting me milk her.

It took a long time for the penny to drop but at last the chain of ominous coincidences began to make their meaning clear and one day I went into the byre with my milking stool.

Disregarding a tetchy look of frustrated ambition in Charlotte's eye I went straight up to her and whispered in her wooshy lug "How is Mr Rochester?" The beam of startled comprehension in that bovine eye was extraordinary. Since then Charlotte and I get on well.

What's in a name? I don't know but I'll take care to call my new beasts by innocent names. Think if I'd called one Lady Macbeth or Clytemnestra.

HORSE PLAY

2 December 1961

We were haring along in a bus through the Christmas card scenery of Elgin's first tentative snowfall when we saw two riding ponies sniffing the winter air in a field beside the oak wood. The sight was so astonishing to eyes accustomed to nothing but tractors in the landscape that we all rubbed circles in the steamed up windows to gaze at creatures we thought lived only in circuses.

Moray must have been one of the first counties in Scotland to mechanise its farming and this for no other reason than that horses in these parts are peculiarly susceptible to grass sickness – a disease as mysterious as fatal.

Grim necessity drove farmers to the tractor for, whatever teething troubles these early implements had, at least a bloke from the garage with a bunch of tools could always cure the upset; a circumstance mightily comforting to a man who had lost an expensive pair of horses earlier on.

For years now Moravians have taken it for granted that horses had no place in their rural economy. Now the new game of pony-owning and pony-trekking has come along and horses are beginning to be seen once more in our fields.

When I was young and lived in the central Highlands, where everyone was very sporting-estate minded, there was a great and exciting trade done in the Highland garron business. Game-keepers, hoteliers, and horse copers joined exuberantly in the fun. One and all expressed themselves knowledgeably about the riding ponies for the toffs; pannier ponies for the grouse; deer ponies who could see their way home in the dark, over the zig zags of all the mountain tracks from Ben Alder to Ardverikie.

Gillies stumbling behind them at the dark end of a long day's stalk would tie hankies to the horses' tails rather like a rear light, and trusting to their surefooted guides would reach their bothy and the larder beneath the flare of the last September stars.

Changed social circumstances following the second war put an end to exclusive sport in the Highlands. But a people with a passion for horse-trading were not so easily discouraged. Along with the breeding of collie dogs, who were certain to take every silver cup and trophy in Badenoch, pony-trading with the hawk-nosed pirates out of Perthshire was the noblest hope of gaining fame and money.

So I was not very surprised to see one of the great specialists in the business turning his unquestionable skill from the letting of the pannier pony to the leasing of the trekking pony. If every one in Britain does not know by this time of the delights and thrills of pony-trekking in the rugged Scottish Highlands it is not because imaginative advertising has not done its most picturesque. It was interesting further to note that pony-trekking is a useful character builder.

There you are, towering crags, rushing streams, a noble steed, and your morals being elevated every minute while the great playground of the Cairngorms and the valley of the upper Spey is spread for your delight. Have I left a single organ stop untouched?

But not even the maddest Highlander can sell the pony-trekking game the whole year round. Winter with its ice and tempest must put a term to it. Therefore the owners of the ponies have to look around for quarters for them. Many of them can be wintered at home, but as the business grows and more and more ponies are needed fields usefully situated are in increasing demand.

In addition many parents with a hippophile offspring are being dragooned into buying ponies. One sees flights of little girls all velvet-capped and jodhpured running about our country towns. They look so enchanting that "horse play" ceases to wear its unhappy connotation.

A SMELL OF SNOW

3 December 1960

Yesterday I was cleaning up the dubs from the scullery floor what time the burns, full to the neck with flood water from weeks of incessant rain, ran full tilt down the hillsides, bawling as they went. To-day I am rushing round in the silent frost with a kettle of boiling water thawing out the hens' drinking fountains which the sudden cold snap has turned into solid blocks of ice. The sun is shining on a landscape too long shrouded in indeterminate rain clouds and misty vapour, so we recognise with pleasure the old familiar features of our countryside. Down in the Laich the firth lies blue and emphatic while against a crystalline horizon the hills show all their precise detail under the year's first definite fall of snow. How good after all those days of breathing the sad sogginess of wet plantations to smell the clean sparkle of snow.

Shudder

Not everything however has taken so delightfully to the change. The kale, for instance, cannot withhold a shudder as the raindrops, marooned in the green crinkle of its leaves, turn to icy diamonds while the grass we are strip-grazing looks downright disconcerted. Perhaps at this late date grass is more slobbery than nourishing for the beasts but in a year like this when, owing to the foot-and-mouth standstill, we shall have to keep cattle longer than we had planned, every blade of extra winter keep is precious.

It is curious how the smell of snow in these upland parts has the effect of drawing all living creatures into closer propinquity, not only the farm animals but the wild untamed birds and beasts who live on the moor. Cows and calves begin to bundle now towards the buildings seeking shelter for the night. From now till the middle of February we have far more night than day, and, though our beasts are for the most part outwintered, they appreciate a bield against the long cold dark. We gutted an old cottage to make a shelter for the increasing herd and we left the ancient fireplace at the far gable end.

The fireplace still has its old sooty 'swye' hanging above the black stone hearth and it is comic to see the cows moving in at night to settle down beside the hearth as if for some placid bovine ceilidh. Every cow has her stance and woe betide the impertinent calf who tries to usurp it, but for the most part the old house is a warm, kind, secure place on a bitter winter night, when the fox barks on the bare hill and owls, soft as sleep but a lot more deadly, hunt their pitifully squeaking prey down the long rides in the wood.

Roes, soon to shed their dainty antlers, cough at the back of the steading thinking of the damage they can do in young growing plantings in spite of the never-ending feud that flames between them and the forester and keeper. Mountain hares wear new white fur coats and, driven by hunger, raid the turnip and kale parks, while little weasels "squeak and gibber" at the door behind which new baby chicks sleep under their infra lamp.

In residence

The wren who comes every winter to the ruined henhouse is now in residence. She is a solitary tiny thing no bigger than my thumb and in spite of legend has nothing to do with the rumbustious robin who lives round the corner in the neuk above the porch at the backdoor. Our two blackbirds natter melodiously to one another about the greediness of the sparrows at the hens' trough while an impossibly decorative magpie pretends he is a lot more frightened than he is. For ourselves the smell of snow sends us scuttling to the circular saw to get the last load of birch cut up into suitable lengths, while a neighbour anxious to get the best out of this season's peats, cleans his chimney by setting it on fire.

223

RIDING IT OUT

9 December 1961

It has snowed since yesterday midday, and now at 10 o'clock the next morning we find ourselves looking on wastes than which Siberia was never colder or whiter.

Now and again an inquiring friend rings up from Forres to say they are knee deep in slush down there but they suppose we have gone all Eskimo. Nor are they wrong, quite.

Certainly the huddle of the house and steading do look like an igloo encampment; but the black patch on the byre slates reminds us that, whatever is to do in other parts of the frozen north, on the outskirts of Dava Moor there are herds of hungry cattle demanding attention now.

We were relieved to find the entrance to the silage pit free from drifting, so the bigger beasts were able to go along and help themselves to breakfast under their own steam; but to hear them you'd imagine they had not seen food for weeks. Actually they are not bawling for a meal but for their calves, which we have shut away for their own good. But the cows think there is no substitute for motherly love even if it takes the calves up to their stomachs in snow, with the cold striking right into their poor little livers.

This storm has taken a day or two to come to culmination. First we had a powdering of snow, then frost which was immediately succeeded by such a brilliant burst of sunshine that we imagined we'd gone all Swiss. Nor were we the only ones to be deceived. Cars began to purr up from the lowlands crowded with ski-ers. We caught glimpses of gay anoraks and bright woollies under the car roofs so securely tied with skis.

It began to look as if from Grantown to Aviemore there would be a wonderful flowering of winter sport. And then the wind rose out of a particularly menacing firth where infernal coloured mountains brooded above gurly water; and with the wind came the snow.

The telephone wires which had been thrilling with gay inquiries about tourist accommodation and the state of the ski-lift suddenly changed their tune, and now began to jangle with anxious questions about whether or not the school bus could make its round in the more isolated parts of the district. The draconian closing of country schools in the North-East now shows up as not such a convenient measure after all, since winter weather in Scotland can make a silly fool of the neatest, tightest schedule ever devised in the comfortable office, all filing system and central heating.

Nor are matters eased by the fact that the propaganda aimed at keeping children longer at school is bearing abundant but distinctly wersh fruit. In the old days country parents did not particularly care whether their offspring lost the odd day from their lessons. To-day, with the furthest depths of Scotland aflame with 11-plus consciousness, every home is sure that absence from school will do irreparable harm to the pupils' future.

Soothing noises cut no ice – literally or otherwise – so haggard roads departments have to recognise the bitter fact that unclassified roads may have to have equal priority in snowplough and sanding with the main highways. Speaking of snowploughs, we have just had a message from our neighbour down the road to say he has a snowplough sitting boiling at his front entrance. Unable to proceed either up or down the brae it stands exhausted, steaming its poor heart away.

For ourselves we are thankful we have the wintering sheep in by and that we have hay for them. We have no desire to venture into a sellers' market for this commodity. If we knew where the two loads of draff we have been promised since a fortnight ago were hiding we'd be even better pleased.

MOON, SKY AND STARS

10 December 1945

We saw the young moon yesterday night on our way to the byre, and nodded to her delightfully, for according to her presence or absence do we count our visitors. She is ever our lantern in these country ways, and when she is gone we stay immured by our own firesides, not daring to seek one another's company over our disastrously bad farm roads.

Speaking for myself, I am not so fond of the moon (except for social purposes) as I am of some less flamboyant astral bodies. Diana is too much the mistress of theatrical effect to commend herself entirely to my ageing heart. Her orange chariot rises lightly as a bubble over the low hills, and mounting swiftly turns to silver as she rides the marches of the night. It is all so bland, competent and serene.

Sometimes though she can disarm me. When "the thin grey cloud is spread on high" and a wind sprung from the cold vexes the sky, then she can shudder like a sick woman in a shawl; and as the watcher looks, she sows her withered light in the desolate fields.

Then the edges of the world become blurred, and but a step separates us from the unknown lands beyond cardis confines. Then are we made conscious of our twin heritage of finite and beloved earth, and unimaginable eternity.

Mostly, however, I prefer the light or stars. Let there be frost to clear away the mirk, and then one can walk at ease under the great dark dome of heaven with, for company, the immortals apparelled in their own celestial light.

Who would not turn with Castor and Pollux in these December nights to watch the Milky Way divide us from Orion; and who, seeing the square of Pegasus, would not rejoice for brave deeds done and lovely stories told?

The very names – Cassiopeia, the Pleidaes, Leda – are like the chords in some old heroic song. Orion swings away to the southeast forever resplendent, and the patient Plough in the north beckons me to fresh labour and blesses what has already been accomplished.

Some nights there are no stars, nor moon either. Then the obscurity is startled by wild fire. It comes like the wakening pulse of the still embryo storm. On such nights witches are abroad, and humble human beings count up whether there are enough stores home before the tempest breaks, and devoutly hope the stick shed is fuller than it looks.

But storm lacks not harbingers more majestic than errant wildfire. As winter draws on, the great space between the zenith and the sea becomes illumined and made glorious by the fallen children of light.

The Aurora Borealis, Northern Streamers, Merry Dancers – call it what you will – flames in the northern horizon, now flickers, now dies, and then rekindles itself. Spectral colours suffuse the mountains and the still firth shines like a fallen shield.

At last the insubstantial carnival is done, and the dazzled watcher feels his blood congealing, and turns himself home at last to his expectant hearth.

227

HOT WATER, ELECTRICITY, AND VANS

10 December 1960

This morning the roads are glittering with ice, so that we seem to be living in an illustration from Andersen's "Snow Queen." Our reaction to all the Arctic splendour of swooping hills and radiant sky is regrettably practical, for with never a glance at shining peak nor beckoning distance, we make with a shovel for the nearest gravel heap to scatter sand at the strategic entrance to farm and field. We have no intention of being frozen to immobility, still less do we want our vans to be stuck on the brae or glissaded into a ditch.

Shopping here is done mainly through the vans, which, if you are the organising kind, you can have calling every day, including Sunday, that is if you buy a Sunday paper. We find such service extremely convenient, and if we do not perhaps discover on the van the same variety as in the shop, we put against that the obliging nature of our vanmen.

Lifts and News

Tom, who brings the groceries from Grantown, also buys me fish as he passes the fishmonger's. He gives neighbours lifts, and keeps us posted in what shop stocks what new line in town. George, the butcher, carries fresh buns to me and an ironmonger's parcel to a friend. Alec will take a prescription to the chemists and the electric fence to the engineer.

No wonder that country women thinking of amenity put the visit of the vans high on their priority list. Hot water, electricity and vans would, I believe be a fair ranking, and if you wonder about the vans I must remind you that not everybody in the country owns a car to run to town when they forget the salt or matches. If in addition to the aforementioned advantages you also possess a fridge, you are indeed sitting pretty, especially if you are a farmer's wife; for now after all the centuries you can pension off the milk cow, that inexorable tie of country life, and having filled up with half a week's milk from the milk marketeers, you can relax till this time three days hence when your milk will be delivered again.

Pedlars

But we have other vans than the food ones. Some come selling carpets and household linen. These we consider chance visitors and not on the same footing as the brush man who comes once a year and knows by instinct when your coir doormat needs renewing and when your furniture polish is done.

We are beginning to count the Indian pedlars who trade in clothes as among our regulars now. They drive along our Highland roads with a degree of Eastern mysticism notably absent when they drive a bargain instead of a car. For myself I never deal with them, because it upsets me when, looking on my white hair, they call me "Mother." By no stretch of the imagination can I think of myself as parent to a nice fat Indian with a beard and turban but Dod loves them and beetles off to make peculiar buys from them. "Flame coloured taffetas" for the girls and a lined helmet for himself that Caligula might have worn on a bashier bash than usual.

No Tinkers

We have come a long way since the war, when the postie was the universal vanman. To-day every shop seems to have a supermarket on wheels. But one firm – or is it tribe? – has refused to take part in the competition. The tinkers have not been here for years. The last time I bought from them was when a basket cost me five shillings and half an hour's acrimonious bargaining. When the tinker lad pocketed my money he sighed and said "Mistress ye've gotten gev auld since the last time I was here." I expect he thinks I'm dead by now.

IT'S NICER TO LIE IN YOUR BED

17 December 1960

It is exactly twenty minutes to nine on a frosty morning in mid-December and the dawn is at last coming rampaging over our low eastern hills. What a splendid noise it makes! Striking great golden chord after chord till the iron hills, sown with frost, resound beneath the arch of high heaven; and the sun, wakened by the fanfare, leaps from his bed to come striding out of Cromdale. The vapour outlining the course of river and burn is thinning and the birch trees, hung with the diamonds of a thousand frozen dews, glitter like enormous chandeliers by the roadside and over the waste places. Pheasants bright as "jewels in an Ethiop's ear" flight among the conifers, while the magpie thinks his black and white as elegant as the blackbird's golden bill or the scarlet combs the grouse pins up above his eyebrows.

Our whole brilliant emphatic world is enclosed within the great crystal bell of upland winter so that we shine half as bright again and sing a pure but heartless song. To be unappreciative of all this magnificence seems sadly ungrateful but the fact is I'd trade the lot for one spring day with daylight eight hours long. Just now we have six, albeit a gorgeous six.

No Virtue

I hate and abominate getting up in the dark and the older I grow the more I hate it. Nothing will persuade my inmost soul that there is virtue in hauling your self and your benighted consciousness out of a nice warm bed and delicious sleep while the Plough has still to finish a long celestial furrow above the larch trees in the west. It is true that I have the benefits of all kinds of gadgets to make the business of rising less painful, but not even the latest in automatic tea-makers can make the inducements of sleep less desirable.

Muttering "Sleep the innocent sleep – balm of hurt minds, great nature's second course, chief nourisher in life's great feast," I stagger to the kitchen switching on electricity as I go, and wondering sourly about the benefits of progress. If this is civilisation, give me the squirrel sleeping soundly in his drey with his nose in his kindly bushy tail and a handy nut in the cupboard.

Silent Hens

What is this life that sends me out beneath the cold bright moon to fill hens' mash into jangling pails while the owl still hunts above the frosted woodland tracks? In their houses the hens are still silent and asleep, save when a downy Sussex utters a drowsy cluck from her foolish dreams. The cats sleep in the warm byres where occasional gentle noises serve to remind me that no one is awake in all the spectral world but me and the hooting owls.

Filling kettles, making porridge, thawing milk basins, I yearn painfully for spring. Demeter never mourned Persephone like this and anyway she had only to await; she did not have to wander the gloomy halls of Dis like I am doing. The cupboards are full of preparations for Christmas and the wood shed is heaped with split birch logs to burn their gay blue flames in our winter hearths, but I remain inconsolable, bereft of spring under the cold winter mould. I wonder how many more are like me, devout heathens supplicating at an ancient altar.

Thankful

But as the moon fades to the ash of tissue paper, one recollects with a sharp pang of thankfulness that it will soon be the shortest day, and in fact three days after this is printed the year will be on the turn. Full well I know that till the end of January I shall still be rising with the undefeated stars and that winter storm in lengthening light can be twice as tedious, but I'll go on cheating that spring is nearly there even to wangling the supper hour on the 10th of February so as to say "Spring is here. Look, we're having supper in daylight."

PROGRESS AMONG THE PAGANS

24 December 1960

When I was young, nearly 50 years ago, Christmas in the Scottish country was never very much of a festival, for the obstinate pagan in the Scottish heart clung with exuberant tenacity to the celebrations at Hogmanay and Ne'er Day. Therefore, though I recollect Christmas as a delightful and even exciting day, it was never regarded by the community as anything but a pale foreshadowing of the massive junketing held at the end of the year.

Cautious

Nor do I find any remembrance of Christmas as a specifically religious time. We did not go to church on Christmas Day unless the date happened to coincide with the sabbath, and in that case we allowed ourselves a cautious recognition of the origin of our faith by singing a Christmas hymn. "Oh, come all ye faithful." Having sung this with melodious fervour, for we were naturally a music-loving village, we wiped our nervous Presbyterian brows lest we might already be teetering on the brink of Mariolatry.

We have come a long mellowing way in the past forty years and though, in the country at least, we still refuse to go to church on Christmas Day we look forward to singing carols on the Sunday most convenient to that date. We send the local bus round our wide district to collect anyone who has not his own means of transport, and we continually express our astonishment in meeting again folk whom we have not encountered since the same time last year. Discreet garlands of evergreens in the vestibule and a pot of flowering hyacinths on the baptismal font testify to the fact that at long last we find ourselves in step with the rest of Christendom. Having come so far we find with a shock of pleasure that we are now prepared to take part in all the other practices which through the centuries have made themselves inseparable from the time. Of course we have had Christmas trees for a long time but we begin to think it would be rather cheerful to do things like carol-singing round our moonlit upland farms and even a nativity play is a possibility, though we feel this may be going rather far.

Because we are so passionately individualistic we attack all these seasonal delights in our own way. I can think of nowhere else that it would be considered quite natural to celebrate the children's Christmas tree and the birth of our national poet on the same night. What had happened was that this particular winter was especially stormy, so that roads remained snowbound and impassable all through December and on until the last week of January, when the situation eased. The entertainments committee in the glen seized their chance and a boisterous evening of wildest merriment was the result. Gaelic toasts and Christmas carols mingled with "Auld Lang Syne" and "The Deil's awa' wi' the Exciseman" to everyone's delirious joy. Now that evening has passed honourably into the realm of myth and legend.

Words and Music

Some trees are more enjoyable than others. There was one I prefer not to dwell on unduly. Santa insisted that all guests should sing "Away in a Manger" under his precenting. He knew the words but not the music, and we knew the music but not the words, so that for one guest at least the manger could never have been far enough away.

Last year, very cautiously, our local Girl Guides ventured on a round of carol-singing in the parish, an innovation which earned them great admiration and unlimited chocolate biscuits. It is a measure of how far we have moved with the times that now we know that port and sherry are alcoholic liquors and as such unsuitable for Girl Guides. When I was young only whisky and brandy were reckoned alcoholic.

I mentioned earlier that the glen was thinking of having a shot at a nativity play and though we admired the courage we were a little alarmed at such a modern notion. No one listened to my small murmur that nativity plays were among the oldest plays we have. Anyway, as I am only a spectator what do I know, but if I could give a general idea of what a Roman solider wore it would be helpful. I should be less than human if I did not speculate on how Hillies's tractorman is going to look in a cardboard helmet and frilly skirt.

Am I being unduly irreverent, too, if I wonder how some of my nice sonsy friends are going to look as eastern kings and queens? But at least it appears as if I am to have a merry Christmas and, oh, I do hope you have the same.

THAT AWKWARD WEEK BETWEEN

30 December 1961

The week between Christmas and Ne'erday lies awkwardly on the Scottish calendar and conscience since it emphasises for us more clearly than we think necessary, the chasm separating our Christianity from our native heathenishness.

We find it difficult to reconcile the grace and holiness of the Rose of the World with the comfortable dark gods whose shrines are the marching forests of our northern latitudes as well as the more riotous enclaves of the Scottish souls. However, we tell one another firmly that Christmas is for the children; New Year's Day is for the grown-ups.

We put past the week between the festivals in extrovert uproar which in the country takes the form of shoots. The air over the moor is clear and cold, as full of sunshine as frost, and parties of shooters bang away cheerfully all round us. There are foxes in the plantings surrounding our farm, and a day or two ago the keepers and locals organised a fox hunt. Fox hunts as conducted in upland Scotland must be the noisiest sport on earth.

For one thing they begin with a loud meeting of Land-Rovers, all driven by men who hold the wheel as if it were a stormy element rather than a piece of precision tooling.

After passionate argument, in which the last draughts match between here and Nairn gets much adverse publicity, each party divides from the parent body and departs to its appointed post. The moment they reach it clamour breaks out in earnest.

The nature of their work has given country people the ability to converse amiably with one another across a 12-acre field. Thus we have developed resonant voices for normal communications. When we begin to shout we really are something. Invigorated by the clear air, made jubilant by the imminent prospect of Hogmanay, the hunters bellow gloriously from crag to corrie.

The fox, whose tracks were clearly visible in the morning snow before the wheel marks obliterated them, digs himself farther into his den, congratulating himself the while on the fact that he had early breakfast off a mountain hare while the moon was still shining on the sparkling lawns of snow that lie athwart every rolling hill.

We do not confine ourselves to fox hunts. We have hare hunts and roe hunts, and bring to all these forays the same jovial racket of racing engines, gurgling whisky flasks, and banging guns. Our text for the week, in short, seems to be, "Without the shedding of blood there is no remission of sin." And then in a bleeze we forget the lot. We go to the village hall, and, holding one another's hands, we sing "Auld Lang Syne."

HAY AND KALE ON THE HOGMANAY MENU

31 December 1960

To-day is the last day of the year, Hogmanay, a feast rich in memory and tradition to every Scot. No one knows the exact derivation of the word but I like to think it comes from the ancient Latin refrain "hoc in anno" (in this year). Such an origin leaves on the ear the echo of a song which surely is appropriate to a nation, one of whose gifts to the world is the international national anthem "Auld Lang Syne."

With all that Hogmanay implies in revelry and festival it is rather downtaking that on an upland hill farm, where the business of rearing cattle is the main preoccupation, it should be a day just like any other. Forty bawling cows and forty bawling calves insist that this must be so. There is nothing so sharpening to the appetite, they aver, as long sparkling nights beneath the frozen stars. Breakfast must be not only ample but up to time, and all other meals are expected to be just as punctual. If there is to be any feasting, the beasts are to have first place at the table.

Plenty

For convenience we have divided the herd into several lots and each contingent has its own bill of fare. I know hay and kale do not sound as convivial as black bun and shortbread, but the dry cows and stirks consider it a Lucullan repast. All they demand is that there should be plenty of both. After they have fed they go stringing out to the hill prospecting for suitable ski-ing slopes, for the moor is frozen hard beneath a powdering of snow and the gay hard sun invites all to winter sports. It is easy enough to provide a suitable Hogmanay for these beasts and even Wee Hughie whose nature does not mellow with the passing months has no complaint.

The herd at home are a different matter. Cows feeding calves are great demanders, and when they are feeding more than one calf they are even more vociferous. And now variety is necessary. Of course they like hay but it better be the good hay and no excuse about last year's wicked hay weather is valid.

Impatient

Old Morag who came here some years ago from the Isle of Skye is especially snooty about second-class hay. When she arrived she looked as if nothing but heather, and very unnourishing heather at that, had passed her lips. To-day no cow in the byre is more impatient of a falter in attention. The hay hardly passes into her first stomach than she is demanding her second course of tatties for the first gallon of milk; and if we think we are to get cream for our pudding on top of that, we better give her a ration of dairy nuts. If they are done, she will make do with the calves' cakelets.

Dod runs subserviently at her bidding and apologises abjectly for the position of her drinking bowl which is too high for her to drink from without her standing on her tiptoes. Muttering Gaelic swears she says her feet are killing her as she splays her way to the open trough over the frozen ruts in the court.

So the day passes like all other winter days in these uplands, to the slow rhythm of the beasts. Four o'clock sees the first pricking of lights in all the other steadings in the glen where our neighbours are doing exactly the same as us. Not till the last hungry cow is fed and the last butting calf has had its supper will any of us be at liberty to have our own meal and then get dressed for the whist and dance which has always been our way here of welcoming in another year and seeing old friends. Like the god for whom the new month is named, we simultaneously look forward and back, and linking hands sing "Auld Lang Syne" before toasting the New Year.

SNOWSTORM STANDSTILL

January 1946

Well, the storm came at last, and so to-day we are knee-deep in snow, with everything frozen into glittering and implacable purity. After the beasts are done our work is over and there is nothing we can do in this weather. We might, it is true, cart dung but the tractor paraffin is finished and we won't get more till to-morrow. So here we are with a whole bright day of enforced leisure.

I am sorry to say that nobody but myself seems to enjoy it. The older one grows, the more gracefully one gives into the inevitable. So it was with a tranquil mind that I lit the sitting room fire earlier than usual, and settled down to an orgy of last month's farming papers and a general cleaning up of letters.

Smithy rampaged home from her calves and said grimly that she was going to clean the cupboards. She's still at it, aided by an indefatigable two-year-old. For shame's sake I put in an occasional appearance and murmur, "Relax, my infants," but in vain.

Walter, I know, will take the gun and career over the fields looking for rabbits. I know so well where he will do so. Down by the now cultivated stubble and over the silent burn. The snow is crisp and hard and the rabbits' feet leave no trace on its surface. The spaniel, black as an exclamation mark, is snuffing at his heels. A few last rose hips, red as drops of blood, still linger among the purple briars. Birds ate all the rowans long since, and the birches dream amidst their smoky twigs.

My own fire burns gaily, with logs giving up their countless summers' suns in one last sacrifice. Where the frost has melted from the window pane I can see the outside world, gay and heartless as a fairy tail. Our Italian is coming towards the house and I pray fervently that there will not be a stern demand for the Signora (that's me).

Frank has a passion for tidiness which nobody else shares, and to-day he is spending his spare-time redding up the farm tools and putting them into neat little boxes. Other days I see him at meal-times, when he treats me to an oration on the shocking way I keep my implements.

To-day I really do not want to hear that the spare spade lugs for the tractor are knocking about the turnip shed instead of being in their appointed places in Frank's boxes. And further, I do not wish to discuss the whereabouts of three new bag needles. Lastly, I do not wish to hear that if I do not mend my ways I shall go bankrupt.

This is my day. It's not often I have a holiday.

LIKE THE EGGS AT GRANNY'S

21 January 1961

We went out of commercial egg marketing last year because we found it did not pay so well as our beef policy. We turned the deep litter house back to its original purpose of a cattle fold and we sold all our hens. Then we bought a couple of dozen young hens whom we housed in wooden henhouses near the wood and kept on the free range system.

All we asked of them was that they should lay us sufficient eggs for our use; and this they have done in an industrious and exemplary fashion. Not only have they laid us eggs but what delicious eggs. The whites are firm, and quite free of that slobber which can be a characteristic of eggs laid by year-old hens in deep litter. The yolks are a deep rich gold and the flavour has that blend of creamy crispness which we are apt to think belonged only to those eggs we got as bairns from our granny's croft.

We are delighted and though we'd been previously eating deep litter eggs for many a moon, when I bought a few litter eggs to help out Christmas appetites I was assailed with disgusted abuse.

Choosy

Now we are not the only people who are choosy about eggs. I go shopping in Forres and find the most unlikely shops doing a trade in "farm" eggs. It looks as if every shopkeeper who has an auntie in the country can develop a trade in eggs. He has a list of eager customers who say they do not mind "paying extra for a fresh egg."

Comparing prices with those for the stamped eggs, I have been interested to note how very much extra people will pay for what they fancy. It is not part of my business to tell the egg packers theirs, except to comment that however efficient they are, eggs they collect can be a week old on the day they collected. There does seem a case for inquiry, not t so much into costs as into what folk are willing to pay for quality.

Make no mistake about it, eggs produced either by the battery or deep litter systems are produced efficiently and cheaply. Eggs from free range call for time wasting devices like hot mashes and warm drinks before breakfast. They pre-suppose little extras like droppies of skim milk and household scrapes like bacon rind and anything else rich in protein. Instead of the henwife going into a warm litter house with an even temperature where the light switches itself off and on automatically while she weighs out an exact ration of pellets into scientifically placed feeding troughs, on free range she must emulate the hens and poadle about outside in all weathers, many of them nasty and often stormy. She must not mind doing menial jobs like thawing out drinking fountains in the kitchen sink and scraping snow out of

243

troughs with a handy little shovel. Her reward? Many, many fewer eggs than from either of the housed systems but of incomparably higher quality.

Higher Price

I am sure there is a market for such eggs and customers will not grudge the higher price. If only economists and such like clever blokes would stop yapping about the necessity of cheap food for an industrial nation, there would be an increasing dèmand for quality foods that taste like they should taste. How many "Herald" housewives have been asking pertinent questions about Golden Wonders – the quality potato? Gold Wonders are not cheap to produce but people want them and will pay for them. The affluent society can afford a palate.

What about broilers? I'm a little tired of the broiler boys saying how wonderful they are and how they have made their industry pay and all without a subsidy. Nobody says what the broilers taste like. I bought one ready roasted for the table at half a guinea. After I'd made a huge salad and boiled a lot of eggs it made a supper for four and tasted like tender flannel. A properly matured chicken fed on milk would have set me back twice as much as the broiler but which would you buy?

"THE YEAR'S PLEASANT KING"

25 February 1961

Like everybody else, we have been enjoying a prolonged spell of delightfully spring-like weather. We are the more appreciative of the gentle air and tremulous sunshine because usually in February we are sliding about in cold slush, feeling every storm doubly long because we cannot now light the lamp and draw the curtains against the snowy landscape. But to-day the distances are blue and there is not a speck of snow to be seen on the mountains over the Firth. We feel rather as if we had won a sweepstake; or had been specially singled out by the fortunate gods for favouritism.

Ploughs are busy on every upland place, and, even though lots of us are having serious thoughts about the economics of the traditional cereal harvest in these modern times, the ancient sense of achievement and hope is as exultant and potent as ever. Farm costings are of more interest to our pockets than our hearts. Anyway, if one must talk economics, are we not enjoying a wonderful bonus in the matter of winter keep? As a rule, February finds the wintering sheep hanging about the home fields looking for a hay ration, whereas to-day they are far out on the moor voraciously eating the luxuriant moss-crop which will do them more good than all the meadow hay in Scotland.

Cows and calves too are more easily kept. After their breakfast of silage the herd make their slow way out in the direction of the old croft. When the beasts reach the big L-shaped planting which shelters us from the south, they lie down contentedly to chew the cud and watch the drifting clouds write their shadowy evanescent signatures along the flanks of every undulating hill.

Wind's Music

The sky, translucent and tender as the inverted calyx of some enormous flower, is full of larks singing. The wind blows the happy enchantment of their music into rainbows of melody to span the firmament.

I hear the wind beginning to rise just as I start the morning chores. The first gentle sighs come tranquilly from the little coppice fringed with beech and then as the day advances the wind grows with it till by 10 o'clock it rushes hallooing triumphantly across the hill and up the river valleys. How different wind can sound in different trees. A thick plantation of conifer can mute the wildest gale, while a few great beeches or elms spaced widely in a park can turn Zephyr to Stentor just in the instant.

Wise Birds

Naturally we do not expect this heavenly weather to last. All March is in front of us, and it would not be the first time we found ourselves snowbound and winter drifted in that boisterous month. Because they have good memories and bitter experience the peesweeps are remembering March as well as we and resolutely refuse to be duped into believing Spring means what she says.

Sparrows may inspect the throat of the forage harvester with a view to setting up a nest, and jackdaws may swing on the trapeze of the warm wind uttering their funny liquid squawk, but the plovers running by the salt sand on their reed-thin legs think the time is not yet for coming inland along with the curlews and black-headed gulls. The twenty-second or third of March is time enough to come clapping square wide wings in alarm through the long spring dusks of marginal Scotland. It is curious how invariable their date is. You might almost set your calendar to it.

But for us storm is to-morrow, and we enjoy our present happiness, even if we count it by days. Daffodils blowing in our grudging gardens and the blue plume of early heather burning help us in our self-deceiving.

IT'S QUIET DOWN HERE

27 February 1945

I'm writing this on my knee about midnight after one of these nice quiet country days.

Immediately after breakfast we started in to fill bags with potatoes, picking out an odd frosted one as we went along. John, the Keeper, was helping me, and as we went along we discussed the recent frost and the festivities which had attended the wedding of one of our neighbours.

Suddenly John said, "There's a float away up to your other steading. Maybe you'd better go and see what's up."

"Och, no," I said, comfortably. "It's just the new bull, and there's plenty folk up by to unload him. Besides, I know for a fact that he's an amiable brute." But, alas, for my faith! In a few moments Frank arrived somewhat breathless.

"Mamma mia!" he ejaculated, leaning in an agitated fashion on a corn bin, "but that bull is infuriated. Perhaps he is just tired with the journey; but, oh, he is wicked!"

With horrid visions of a gored Smithy and prostrate babes I shoved a half-filled sack at him and sped home to find Smithy reassuringly amused at the morning's work but agreeing substantially with Frank. Apparently Robert (that's our bull's name) was a bit peeved by his journey and decided to honour his new home with a highly exciting display of fury.

Mercifully the door of his home was open, and in a last burst of rage he flung himself pettishly through it, whereupon the humans thankfully banged the door to and left him to (they hoped) his better nature.

The story wasn't calculated to lull my fears, so I thought I'd better go and see whether Robert had subsided. Very cautiously I opened his door and peeped in. In the next stall to his stands an Irish cow called Molly, who once kicked me from one end of the byre to another. Robert was deep in conversation with this virago. He was telling her of his woes, and she was counselling him not to stand for it. I retired.

On my desk lay a list of the day's 'phone calls. "Memo, ring re threshing mill." "Memo, ring the vet re the neighbour's stot" – which has developed the habit of chewing the hair of the other cattle so that now they look like leopards instead of cattle beasts. "Memo, see about manures." And so on. After dinner Robert was still annoyed, but Frank assured us it was just artistic temperament.

In the evening two English calves arrived by train, having been far too long on the way. So we bedded them and gave them drinks and washed their tails. And now it's midnight and I haven't seen the papers, and, oh me, I'm sleepy!

THE WIND AND THE SNOW

To-day has been one of those days. One of the penalties of living in the country is that one is so completely at the mercy of the weather. It is our constant companion and preoccupation. It is our life, our living, our ruler. Sometimes it is beneficent and then nothing can harm us, sometimes it is our tyrant and then we are undone. To-day it decided to do its worst.

We woke to a girning north wind that overturned a henhouse and tousled the stacks most wickedly. It was bitter cold, and now and then it snowed viciously; small stuff that found its infallible way between scarf and neck. Going to the byre was a penance of wet wellingtons, heavy coats, and lanterns which had been forgotten the night before and were now grimy and spent.

Daylight made a tardy appearance, and as we put out the lamps well after ten o'clock, we spared yet another bitter thought for Summer Time.

The haggard larch trees stood out against the winter sky and bands of mist came and went in the wings of the wind. Occasionally they lifted to show a glimpse of far, wan sea and then they descended again, suffocating the valley and shutting off each croft in its island of solitude.

Showers of hail sent our robin flying querulously round the porch. The dogs haunted the back door fain to lie in front of the kitchen fire, but they were shooed off to the barn.

The men worked in the steading, attending to the beasts, and we caught an occasional glimpse of them with their bodies bent against the wind and a swirl of straw about their shoulders. They came in for a pail of warm water to make a bran mash for a cow newly calved and we banged the door hurriedly behind them.

Mist, hail, and wind, the squadrons of storm raced between us and the hills. The house shivered in its windows and doors. A rattle of machinery told us that the snow plough had gone up and we hoped fervently for open roads. Road blocks to those ten miles away from the nearest town are inconvenient and uncomfortable.

From the woodshed came the comforting whine of a saw, and then the steady clop clop of an axe biting its way through the sticks for the evening fire.

Cows to milk, calves to feed, beasts to straw and water, the day went past, and then it was night. The scullery is full of wet boots, the porch is full of wet clothes, the wind is still howling, and the snow is still nyattering, but at least the day is ended, and we don't have to live it again.

Layout: Stephen M.L. Young
                 latouveilhe@mac.com

Font: Adobe Garamond (11pt)

Copies of this book can be ordered via the Internet:

                 www.librario.com

or from:

                 Librario Publishing Ltd
                 Brough House
                 Milton Brodie
                 Kinloss
                 Moray IV36 2UA
                 Tel /Fax No 01343 850 617